THE Goldfarb CHRONICLES

THE Goldfarb CHRONICLES

MOVING WITH BABY
THE SOLITARIO
BREWSTER COUNTY LAW

WM. HOVEY SMITH

LitPrime Solutions
21250 Hawthorne Blvd
Suite 500, Torrance, CA 90503
www.litprime.com
Phone: 1-800-981-9893

© 2023 Wm. Hovey Smith. All rights reserved.

No part of this book may be reproduced, stored in a retrieval system, or transmitted by any means without the written permission of the author.

Published by LitPrime Solutions   06/26/2023

ISBN: 979-8-88703-195-8(sc)
ISBN: 979-8-88703-196-5(hc)
ISBN: 979-8-88703-197-2(e)

Library of Congress Control Number: 2023904021

Any people depicted in stock imagery provided by iStock are models, and such images are being used for illustrative purposes only.

Certain stock imagery © iStock.

Because of the dynamic nature of the Internet, any web addresses or links contained in this book may have changed since publication and may no longer be valid. The views expressed in this work are solely those of the author and do not necessarily reflect the views of the publisher, and the publisher hereby disclaims any responsibility for them.

Contents

Book 1. Moving With Baby

Chapter 1. Samantha And Aron.................1
Chapter 2. Georgia Grandparents..............14
Chapter 3. Getting To Know You...............25
Chapter 4. Medical Issues....................36
Chapter 5. Working At Home...................44
Chapter 6. Graduation 1......................53
Chapter 7. Beth and Larry....................63
Chapter 8. A Wedding.........................71
Chapter 9. Consequences......................80
Chapter 10. Eats And Meets....................88
Chapter 11. The Wedding.......................97
Chapter 12. Conception........................107
Chapter 13. Aftermath.........................112
Chapter 14. Unexpected Events.................122
Chapter 15. Loading Out.......................132
Chapter 16. Arrow Flight......................140
Chapter 17. Unexpected Passenger..............148
Chapter 18. Marooned..........................156
Chapter 19. Neighbors.........................164
Chapter 20. Sextuplets........................173
Chapter 21. Discovery.........................183

Book 2. The Solitario

Chapter 1.	New Horizons	195
Chapter 2.	Ranch Life	204
Chapter 3.	Survival	215
Chapter 4.	Capture	224
Chapter 5.	Ransom	232
Chapter 6.	Rescue	237
Chapter 7.	Aftermath	248
Chapter 8.	Recovery	256
Chapter 9.	Fiesta	266

Book 3. Brewster County Law

Chapter 1.	Trial Day 1	285
Chapter 2.	Trial Day 2	296
Chapter 3.	Trial Day 3	315
Chapter 4.	Trial Day 4	328
Chapter 5.	Judgement	337
Chapter 6.	Treatment	346
Chapter 7.	Graduation 2	355

THE Goldfarb CHRONICLES

Book 1

Moving With Baby

Chapter 1

Samantha And Aron

STANDING IN LINE AT THE check-in counter for the Delta flight to Atlanta, Samantha and Aron's disparate heights and differences in dress would appear to indicate that these were two business travelers on the same flight instead of a couple going to meet their future relatives. Aron who was taller than his soon-to-be bride by more than a foot was fidgeting with his ticket and checking for the umpteenth time that he was at the correct gate, while Samantha was waiting patiently behind him as might be expected from a Regional Vice President who flew several times a month.

Grabbing Aron's hand she said, "Calm down. This is the right flight, and we are going to get there in plenty of time."

"That's not what I am worrying about. What if they don't like me?" he replied.

"We'll talk about it on the plane. Right now let's just get on board and find our seats. It looks like it is going to be a full flight."

Feeling like a mouse trapped in a cage being taken to his execution, Aron was not much comforted with the prospect of meeting Sam's father and stepmother and her rural grandparents who he imagined were something like characters out of the movie "Deliverance." Even more unfamiliar to his urban upbringing was Sam's step-uncle Tyler who lived next door to the grandparents. Thankfully, Aron thought, Tyler was on a Cape buffalo hunt in Africa.

"What if they don't like Jews?" Aron whispered in Sam's ear.

"That is no problem. They have grown up with Jews. Believe me, they don't care. I see much more anti-Jewish stuff here in Detroit, than I ever saw down there. They eat a lot of pork and bar-b-que, but you do too, even if your parents don't. There won't be any problems. It's you and me that are getting married, not you and them," she concluded decisively.

"Will they come to Detroit for the wedding?" Aron questioned.

"I am sure that mother will, but I don't know about my dad, stepdad and the grandparents. We will have to see. We will invite them, but if they will come or not, I do not know."

Arriving at Hartsfield international in Atlanta they quickly grabbed their carry-ons and made it to the rental car counter where they picked up their late model Toyota sedan and made their way from the airport across Midtown into the city. Samantha expertly snaked the vehicle out of the parking lot onto I-85 to make it to her dad's house on the northwest side of Atlanta in one of the older subdivisions inside the beltway.

"Who are we going to see?" Aron Asked.

"My dad, his wife and my half-sisters will most likely be there. One will be graduating from college next year and the other is finishing high school. We are just going to have a meal somewhere and then drive to my mother and stepdad's house in Covington and spend the night there," Samantha replied reassuringly.

"This is going to be like the worst job interview that I have ever had. I already feel like I am under a hot lamp undergoing a police interrogation, like in those old movies on TV. I will be waiting while they bring out the rubber hoses," Aron said as he ran his finger around his collar to loosen his tie.

"Believe me everything is going to be just fine. They are all curious about you, how you grew up in Detroit and how we are going to live. They know we have been living together and are fine about that. No problem. In fact, Mom was glad that we did, so there would not be any surprises later. They are going to love you just like I do. We are just about there," she concluded.

Off the interstate and driving through streets lined with small businesses, she turned into the Gatepark Subdivision which was much more wooded and parked in front of a two-story brick home with a one-story porch across the front. A driveway beside the house ended at a two-car garage at the back. Concrete slabs had been paved off the driveway to allow for two additional vehicles. Sam pulled into an empty spot and stopped.

"Well here we go," Aron thought as he got out and made his first step onto a patch of red Georgia clay which the grass had not yet covered beside the new slab.

Coming across the lawn from the screen porch behind the house, Tom O'Malley, Samantha's father was followed by his wife, Susan, and their daughters Margaret and Elizabeth. As Aron stood awkwardly by the siblings hugged and Tom walked up to shake his hand.

"I let the women do the hugging," Tom replied as he offered Aron a firm handshake. "Come inside and use the restroom if you need to. I have made reservations at The Abby, an Italian restaurant. I hope that is all right with you."

"I am sure that will be fine," Aron replied. "It is good to meet you. I'm sorry, but where is that bathroom? It has been a long flight and a long drive."

"Follow me," Tom said as he escorted the lanky Aron onto the back porch. "Please take off your shoes. This red clay is difficult to get out of the carpets."

Aron did as he was bid and was escorted through the back door where the kitchen was on the right. Another door led to a large, tiled room containing a hot tub, shower and a pair of stalls containing commodes, such as might be seen in a hotel's fitness room.

"I thought about putting in a sauna," Tom said, "but decided on a larger hot tub that we can all get in together."

As Aron peed, he responded. "Sam and I did not bring any bathing suits."

"That's fine. You won't need them when you stay here. When you are done and washed up, we can leave."

Aron had never felt comfortable about being naked, much less in mixed company, so at least on this point he was pleased that they would be spending the night at Samantha's mother's house.

Tom had reserved a separate dining room at the restaurant so that they would have a chance to talk in relative quiet while the food was being prepared.

"Aron I understand that you work at the University Hospital in Detroit. What do you do?" Tom questioned.

"I work at a section that handles fertility and conception. If a couple has difficulties in conceiving a child, that section does in-vitro fertilization, egg implantation and follows through the entire pregnancy from conception to post-delivery care. I coordinate the insurance coverage and sometimes arrange grants to help couples pay their bills. This is an expensive process, and unless you are wealthy, most people cannot afford it," Aron explained.

"Sam, are you going to be able to continue work if you get pregnant?" Tom queried.

"Although I travel a lot, the majority of what I do can be accomplished by conference calls between the insurance brokers at the company and the company that we are insuring. My company has a maternity leave policy, and Aron's insurance is even better than that. Not only does he get paternity leave, but all of the hospital expenses are covered," Samantha replied.

"That's fantastic," Susan responded. "Most companies are cutting health coverage right and left. For workers with a family the premiums are going up and up every year. To have a child in this day and age is a very expensive proposition, even if there are no complications."

"All of that sounds awful mechanical to me. I think that I would want to do it the regular way," Margaret responded.

"So would we and nearly everyone else, but sometimes that is just not possible. The assisted conception methods and procedures are becoming more precise and predictable every year, and the hospital has saved the lives of many mothers and their infants that would have died in delivery," Aron added.

"You mean women still die during childbirth?" Elizabeth asked with a note of doubt in her voice.

"Childbirth always carries risks. There is no way around it. It is the poor and those who cannot afford hospital care that come in undernourished, with or without drug or alcohol problems and/or psychological issues who are in greatest danger. In the U.S. altogether too many mothers die in medically underserved parts of the country. Having babies is a serious business," Aron concluded.

Thankfully, the meal arrived. Soon there were steaming plates of pastas and sauces along with a baked salmon on the table along with bottles of wine.

Samantha put her hand on Aron's thigh and when he turned whispered in his ear, "You did fine. Enjoy your meal, but don't stuff yourself. We still have a lot to do tonight."

With a round of good-byes, good to meet yous and see you again soons exchanged between them in the restaurant's parking lot, Samantha and Aron were once more on their way. Samantha keyed her mother's address into the car's GPS and proceeded towards I-20 and Covington.

"Mom, we are on our way," she said over the phone before they pulled out of the parking lot. "We should be there in an hour or so, depending on the traffic."

Another two-story house, this time with white siding, greeted the couple when they arrived. They pulled up the concrete driveway into the back of the yard where a basement level with attached screen porch opened up on a grassed slope leading down to a wooded creek valley. To the left and right were similar-sized homes with slightly varied exterior profiles.

As Samantha and Aron were getting out of the car, Fred and Lollie Williams approached along with their German Shorthair, Rover, who immediately proceeded to give the new arrivals sniffs to test the strange scents given off by their shoes and lower legs.

"Rover, get back. Sit." Fred commanded and the dog responded.

Sam reached down and grabbed him behind the both ears and gave him a good scratch, "You remember me, don't you?"

Rover responded with a flop, flop of a tail wag and licked her hand in recognition.

Turning to Aron she reassured, "Don't worry he is a bit shy until he gets to know you."

Aron extended his hand, and Rover bristled up and emitted a low growl.

"It takes him awhile to respond to strange men for some reason." Fred said reassuringly. "Don't worry, He will soon get use to you. Come inside."

"How was your trip down?" Lollie asked.

"Outside of the usual hassles at the airports and the traffic, everything went fine," Samantha replied. "The only thing that was a bit different is the work they are doing around the airport."

"They are going through their fifth expansion I think," Fred added. "They never seem to finish it and, of course, we taxpayers wind up paying for it."

Progressing through the basement up to the main floor to the kitchen and great room, Aron and Fred carried the bags while the women and Rover went ahead.

"There are bedrooms and a bath upstairs. Why don't you two get settled in and come down when you are ready? I know you just ate, but maybe we can have coffee or something," Fred said.

As if to show the way, Rover bounded up the stairs and stood in front of Samantha' bedroom door. When she opened it, the dog jumped into the middle of the bed and wagged its tail, obviously glad to see Samantha home.

Aron looked at the double bed with the dog now occupying the middle of it. When he approached Rover growled and showed teeth as if he was prepared to defend his and Samantha's sleeping area.

"I don't think that this is going to work," Aron responded.

"Actually Rover is my dog. I had it while I was at college. The folks kept it for me when I moved to Detroit. I think you will need to sleep in my sister's room. She teaches Special-Ed in a school district nearer where my Dad lives. So, she won't be staying here," Samantha informed Aron.

"O. K. I'll move my stuff next door," Aron replied as he started to kiss Samantha which solicited a round of barking from Rover. Taking fair warning, Aron turned and went out into the hall where he saw a door with a sign, "Danger Zone Do Not Enter" painted in bold block letters.

"Uh. Sam come here a moment," Aron requested with a hint of nervousness in his voice.

"Oh. That," Sam responded. "Beth never liked to clean up, and kept stuff in piles around the room. Mom got tired of arguing about it and told her, 'If you want to live like that you can, but I am not going to waste my time cleaning up, just to see you mess it up again.' "Consequently, I suspect that her room is just like she left it."

When Sam opened the door, they were greeted by the smell of old pizza, scented cosmetics and stale air. The scent profile immediately attracted Rover who leapt past them to more fully enjoy the pungent aromas emanating from piles of boxes and clothes scattered around the room.

"I'm sorry, Aron, I did not think it would be quite this bad. open the window, and let things air out. I will get some laundry bags and help you sort things out. Take the sheets off the bed, and I'll wash them," Sam said as she went downstairs leaving Rover and Aron to explore their strange environment.

Rover had initial success and clamped his jaws on a pizza box and resisted Aron's efforts to take it from him to the extent that Aron lost his footing and fell backwards across the bed as the delighted dog escaped with his prize.

Assessing his environment for the first time, he saw that the room was painted in a muted rust red with yellow around the window frames, base boards and molding. On the walls were pictures of pastoral scenes from the Romantic period showing classical nude figures engaged in various activities. He removed the pillow slips and found that the pillows smelled so much of women's hair products and perfume that he doubted he could sleep on them. The sheets and bed covering had a fringe that somewhat matched the rust-colored walls. After he placed the bedcoverings in the hall, he turned his attention to boxes of magazines and books beside the bed.

Sam returned with two laundry bags and a large garbage bag.

"Honey, I am going to need help with this. I have gotten use to handling your clothes, but I feel like I am violating your sister's privacy by going through her belongings like this. What are we going to do with these Play Girl magazines?" he asked.

"If Beth did not take them, then she does not need or want them. I would save the hardcover books, but those magazines can go in the trash," she replied.

"Some are sort of interesting like, *The Modern Karma Sutra*, *Fifty Ways to Please Your Lover*, and *The Joy of Sex*. I'll sort those out," Aron volunteered.

"Those are Beth's books, and we should save them," Samantha replied.

Three hours later the room's floor was empty, the bed changed with clean sheets and the room smelled better.

"Thank you two for cleaning Beth's room and washing her things. That has been needed doing," Lollie commented.

"I did not open a drawer and we did not do anything with the closet. Whatever is in those places, I did not want or need to see. Thank goodness Sam helped. I did not feel at all right about handling her things," Aron replied apologetically.

The second round of family interrogations began with the sounds of a washing machine and dryer running in the background, as they began to talk over Bourbon and Cokes. Rover entertained himself tearing up the pizza box he had liberated from Beth's bedroom on the tiled kitchen floor. For the second time today, Aron when through his job description and duties. He was more than happy to pass the baton when Lollie turned to Samantha and asked, "How did you meet?"

"It was at work, sort of. "Samantha started. "I was giving a presentation on my company's insurance plan to the University Hospital and Aron was in the audience. I noticed him immediately because he was the tallest person in the room. Because he works with insurance issues, he had a series of important questions. They became so detailed that I suggested that we meet later when I could more thoroughly answer his concerns and not interrupt the flow of the presentation. Most of the people in the room wanted that meeting to end as quickly as possible."

"I can relate to that," Fred said. "Those corporate and safety meetings at work seemed to go on forever."

"So, I set up a one-on-one meeting with Aron," Sam continued, "and things progressed from lunch dates to getting together when I was in Detroit to our moving in together. There did not seem to be any reason for us to keep separate apartments when we could share one and cut expenses."

Aron interrupted, "My parents helped me buy a third-floor walk-up in an older apartment building which had enough room for us both somewhat near the hospital and airport. That apartment works for us until we can pick out one together."

As Aron and Sam began to yawn and nod off, Fred made the suggestion, "We need to go to bed. We will be leaving at 7:00 in the morning to spend as much time with my parents as possible. We will have lunch down there, visit and come back here."

"I'll go ahead and shower and then you can do the same. I'm too tired to do anything else," Samantha suggested.

"Me too." Aron agreed. "This has been quite a day."

Aron undressed and laid his clothes across a chair. When he heard Samantha leave the bathroom he walked across the hall and started his own shower. As he came out of the shower, the door burst open and an irate woman stormed in.

"Who are you and what have you done to my room?" Beth demanded.

Soaking wet and naked, Aron grabbed for a towel and proceeded to wrap it around himself as he replied, "I'm, I'm Aron, Samantha's boyfriend. We are going to be married."

"Is that so? Stay out of my room," Beth ordered as she went into her room and locked the door.

Drying himself off, he went to Samantha's door and opened it to be greeted by a growl from Rover, which he properly interpreted as a warning to come no further. Returning to the bathroom he found some beach towels in the bottom of the cabinet and he took these down to the living room where he covered himself as comfortably as he could on the sofa before drifting off to sleep.

"What are you doing down here?" Fred asked the next morning as he shook the sleeping figure by the shoulder.

As Aron got his wits together and realized that he was more than a little bit exposed, he covered himself and replied, "Sam's sister, ur Beth, came home and ordered me out of her room and locked the door. I could not get my clothes and Rover would not let me into Samantha's room, so I grabbed some towels and came down here."

"I suppose that Beth has broken-up with her boyfriend again. This has happened before, and until she gets her head on straight, she is not worth living with. She has an almost impossible job as a Special-Ed teacher. She is supposed to design programs for them so that they can meet state-educational milestones that often change. There are educational milestones, social milestones and physical milestones that she is supposed to certify that the kids are meeting. In effect, this is a job that she cannot leave at work, and she brings it home. There is also the 'teacher spouse syndrome' where her default position when she is exhausted is to treat Larry like her students, and this puts great stress on their relationship. No man wants to be ordered about like a child when he makes what Beth considers a mistake or exercises poor judgement. It's like she is keeping a report card on him. I guess something like that has happened again. He will call, they with talk whatever the issue is out, and she will very likely go back to him. Larry is an all-right guy who is an Electrical Engineer who does contract work for a variety of big companies. His job is also stressful with challenging deadlines, and sometimes things just boil over," Fred concluded.

"Can I get my clothes? I can't run around like this," Aron asked.

"You stay down here, I will go up," Fred said and then went up the stairs which was followed by a series of shouts emanating from the stairwell.

Aron caught snatches of "I don't care if he does run around naked after he pawed through my stuff," and "If he wants his bag he can have it," which was followed by a crash outside on the sidewalk as his

THE GOLDFARB CHRONICLES

suitcase landed and burst open. Carefully opening the front door he found that the bottle of celebratory Champaign that he brought had broken and was spewing like a geyser mixing with a broken bottle of aftershave which was soaking his clothes.

Hearing the commotion, Lollie, dressed in her nightclothes and a bathrobe, joined Aron as they watched the spectacle.

"What happened?" Lollie asked.

"Beth came home last night," Aron explained. "She was furious about what Sam and I did to her room and she locked me out. I spent the night down here. Fred went up to get my bag and Beth threw it out of the window. Everything is soaked, and we are supposed to leave in an hour."

"Don't worry. I'll put everything in the wash, but that is going to take two hours which is when we are expected in Statesboro. Maybe Fred can find something for you to wear," Lollie offered reassuringly.

Joining his wife and Aron by the door to witness the dying gasps of the erupting Champaign fountain, Fred said, "I have a set of camo coveralls that I picked up at a dumpster that will fit you. They were ripped, and I sewed a patch on them to use when I paint. There is a retired ex-wrestler who lives two doors down who is about your size who might have some things that you can wear. He runs with Rover this time of morning. In fact, here he comes."

Gaylord, a tall athletic-looking black man with a sprinkling of white in his black hair came trotting down the street. He was dressed in shorts, a T-shirt and running shoes and carried a dog leash in his hands. He responded to Fred's call and wave.

"Gaylord this is Aron, Samantha's boyfriend, and through a series of events that I'll tell you about later he needs a change of clothes. Do you have something he can wear?" Fred inquired.

"Probably. When I was wrestling, I supported a lot of liberal causes and when I made donations, I was often given things that I have never worn," Looking the bare-chested Aron like a potential opponent in the ring, he continued. "You and I have about the same height and reach. If you ever want to learn how to wrestle or box, I would like to work out with you."

Uncomfortable at this unexpected man-to-man evaluation Aron replied, "Right now I just need some clothes."

"Give me about 15 minutes to pull some things out and come over. What sized shoes do you wear?" Gaylord inquired.

"I wear a size 13, in most shoes," Aron responded.

"Good, when I worked at City Parks and Recreation Department, I bought bowling shoes to lend to kids who could not afford them. When I retired, I took a set in case some kids might want me to give them instructions. I'll go and get those things out now," Gaylord concluded.

Rover recognized his running buddy, woofed and greeted Gaylord with licks and enjoyed a good back scratch. "Sorry Rover, not today. I've got to get your new human some clothes. You need to stay."

Rover sat and whined as he saw Gaylord walk back up the street. Then he turned and reentered the house.

As Lollie cleaned up the mess and sorted Aron's clothes prior to washing them, Aron went into the downstairs bath and put on the overalls and a pair of large boot socks to walk down the block to Gaylord's house.

"Is this guy gay?" Aron asked Fred as he prepared to leave.

"That is a persona that he adopted as a gimmick when he wrestled. He was married and had two or three kids I think. They are all grown and moved away. His wife died about three years ago. I don't know what his real name is. Everybody knows him by Gaylord."

Attired in socks and the patched Mossy Oak coveralls, Aron was glad that he was not in a neighborhood where anyone knew him. His only protection for being picked up for a vagrant were his IDs that were in his billfold that was being aired-out.

Gaylord, still dressed in his running outfit, invited him in. "I put some things out for you in that bathroom. Try them on and see if they fit."

What Aron found was a selection of items from the Black Power and Gay Pride movement of the 1960s. These included Act Up printed boxers, Malcom X T-shirt, Martin Luther King shirt, prison stripped trousers and a purple jacket with a yellow triangle labeled Lambda Legal Defense.

When he walked out he felt better about having some reasonably good-fitting clothes, even if they might not be his first choice in fashion.

"These are things that I sometimes wore in the ring to whip up a crowd," Gaylord explained. "You might get some curious looks from the brothers, but I don't know what any white folks might think."

Aron did not know either. He could imagine tomorrow's headlines "White Gay Activist Lynched in Central Georgia."

Chapter 2

Georgia Grandparents

"WHAT THE HELL HAPPENED BETWEEN you and Beth last night, and where in the world did you get those clothes?" Samantha demanded at the breakfast table while Lollie was dishing up scrambled eggs, bacon, grits and hot biscuits.

"I don't know how to even start to tell you. This is going to take a while for me to sort out, much less tell anyone," Aron responded.

"Too right, to use an Australian expression that seems somehow appropriate," Fred replied. "We need to eat and leave, and let Beth think things through."

"All of that shouting woke me up. I talked to Beth, and she told me that she and Larry had brokenup again, and she started to list all of the things that he had done wrong and why she was mad at him for the umpteenth time and on and on. I guess it was good that she got that out of her system. After she calmed down a little, she asked me about you. She said she was sorry that you happened to catch her at a terrible time. Believe me Aron, she is really a nice person. It is just that those pent-up frustrations have to breakout from time to time. I am sorry that you happened to be the object of them. You will like her once you get to know her," Samantha assured her husband-to-be.

"I don't know," Aron responded. "I see her like some demon Viking war-woman up there ready to hack me up with a battle axe. I may have to take Gaylord up on his offer to teach me boxing."

"Enough talk," Fred interjected, "It is time we were leaving. By the time we get back tonight everything will be all right."

"With Beth back, maybe Sam and I should sleep at Sam's dad's tonight. We were sort of invited," Aron volunteered.

"That might work," Fred responded. "Lollie see if you can set things up when we get in the car."

Wanting to change the subject, Lollie asked Aron, "How did you like your grits?"

"I had heard about grits, but never had them," Aron replied. "I can eat them, but the texture is a bit strange, and they don't taste like much to be honest. I was brought up with hash browns for breakfast or even stale pizza when I was in college. That was the students' survival food."

"Next time I will make you some cheese grits or shrimp and grits which have more flavor and a smoother texture," Lollie assured her breakfast guest.

Loading up the SUV, Fred and Aron sat in the front seat, Samantha and Lollie in the rear while Rover hopped up on a pallet in the rear, eager to go for a ride. Samantha rolled down her window so he could stick his head out and check the scents until the vehicle got up to highway speed.

After leaving a half-hour later than expected, Fred drove and explained some of the sights to Aron. "We'll pass Stone Mountain on the right which has the carvings of the Confederate Generals on it. There is a nice park there where they hold concerts and other public events. It's a big blob of granite. You can walk up one side, but it is so slick in other places that people have tried to walk down it and could not get back up. Some have fallen to their deaths on the mountain. Now there is a cable car, like at a ski resort, to take visitors to the restaurant and gift shop. When you come back we will visit it and the Cyclorama at Grant Park which they refurbished a few years ago. There were once others like it at major Civil War battlefields, but I think that this is the last one left and certainly the best known," he asserted.

Samantha was thinking, "Too much information." Although Aron seemed to be handling this deep immersion in southern culture all right she was concerned that all of this might be an overload.

Leaning over the front seat she asked Aron, "How are you doing?"

He replied with a non-committal "All right." This statement did not nearly express his feelings that he was being put in triple jeopardy. First there was meeting Sam's mother and step-father, then Sam's father's parents, and now Sam's stepfather's parents.

"What are your parents like?" he asked Fred.

"Dad, Robert, was a Korean War vet. He met my mom, Charlotte, when she was a college student in Statesboro, and they got married. He went to work for Mede Packaging here in Atlanta and retired from there. While he worked, he and my Mom lived in Covington. After he retired, they purchased a modular home and moved in next to my Uncle Tyler on the family land in Statesboro. There they could get away from Atlanta, he could hunt on the land and this was a less expensive place to live."

"They get along?," Aron asked.

"Tyler's wife died several years ago, and they helped him through that. He lives alone and mostly writes and keeps to himself. He writes for outdoor magazines, and is in South Africa at the moment. I don't know if he makes much of a living at it, but it keeps him busy. His wife was a nice lady who had adult children by the time they married. Her children don't have much of a relationship with him now that their mother has died."

"He did not remarry?" Aron questioned.

"No. His wife, Thresa, had a long, lingering illness – pancreatic cancer. After going through that he decided that he did not want to be involved with yet another family and their relations and problems, etc. He preferred to live his life, do his thing, keep his dogs and live by himself."

"His marriage was O.K.?"

Samantha hesitated to answer, and Lollie took up the narrative. "They got along very well. For their fifth wedding anniversary they went to Paris, and for their tenth they went to Rome. Tyler had been to Paris covering an international conference for one of his books, and he enjoyed showing his wife the city. She was a religious person and interested in seeing the Vatican and religious sites, and those trips were some of the highlights of their marriage."

"I would like to go to Italy for our honeymoon. I have plenty of airmiles and hotel credits so the trip would not cost us hardly anything," Samantha added.

"That would certainly be interesting," Aron replied. "I have never been to Europe."

"It could not be any worse that what I am going through now," he thought.

"We are coming up to where we get off I-20." Fred interrupted. "There is a Flying J there. I usually stop, walk Rover and give everyone a bathroom break. If you are thirsty get something to drink, but Mom is going to have a meal for us when we get there in about an hour."

Dressed in his purple coat, prison-striped pants and Martin Luther King shirt Aron got some stares as he walked in among the tourists and truckers. He quickly located the bathroom and went inside to use the urinal. As his pants had no fly he had to untie the waistband and lower the trousers to pee.

"Man. Where did you get those threads?" a young black man asked.

"They were given to me," Aron replied.

"Who would give a white guy clothes like that? Don't you know that wearing that stuff down here can get you killed. If the rednecks don't get you some brothers would just for the clothes," he informed the startled Aron who was trying to complete his bodily function as fast as possible and get the hell out of there.

"Gaylord gave them to me," he responded not knowing what else to say.

"Man. You know Gaylord! He is my all-time hero. My hero man. When he wrestled, he would walk into the ring with all of that black power and gay pride stuff on and just beat the crap out of his white opponents. It was all playacting and everyone knew it, but that was a fine show. Everybody loves him down here. Can I touch that jacket?" he asked.

"Sure," Aron replied and felt a hand lay across his shoulder and then instantly be withdrawn as if repelled by an electric shock.

"That is just too cool man. Too cool. But if you are going to wear those things let me show you how. Turn around."

Aron finished and pulled his pants up and turned to face is unexpected admirer who pulled the pants down below his butt, passed the draw strings over the tops of the hip bones and tied them with a bow over his belly button. Then he knelt and folded up the bottom of the trousers to make cuffs over the top of the shoes.

A flush was heard from inside a stall and out came a 300-pound bearded truck driver wearing soiled jeans and a Drive-Um-Wild T-shirt with silver lightning bolts running diagonally across the front and back.

"What an outfit. I don't wonder that friend here thought you might need some help. I use to do some wrestling, and although I never wrestled Gaylord, I had seen him. He was a great wrestler and a good man. Come out with us. No one is going to bother you around here," the trucker assured the somewhat startled Aron.

With Aron in the middle, the trucker on his right and his newfound black acquaintance on his left they approached the SUV. Fred and Rover were already inside. As they stood by the door the two shook hands with Aron and the trucker turned to Fred and said, "Look after this guy, he needs help."

Before getting into the vehicle Aron readjusted his trousers and then sat to roll down the cuffs.

"Who were your friends?" Fred asked.

"Just some guys I met in the bathroom," Aron replied. "They knew Gaylord and thought I might need some help getting out of there. They said I could get killed wearing clothes like this."

"A few years ago, a white man was found off I-20 near Savannah. He had been stripped naked and beaten so badly that he had complete amnesia. He never did, so far as I know, find out who he was or what happened. They wrote that story up in the Reader's Digest. I suppose that this incident was what they were talking about," Fred explained.

"I don't want to tell the women about this," Aron requested. "But

I don't want to go out in public in these clothes if I can help it. I am worried about what your parents might think."

"Don't. Here come the women. Let's load up and get gone."

"I had a good signal, so I phoned Sam's dad while we were inside," Lollie informed the pair. "They are all right if you spend the night. I explained that you might be arriving a little late, and Tom said that any time before midnight would be fine. He said that he would have the hot tub ready so you could relax."

"I think that I will pass on that one," Aron responded.

"Don't be such a prude," Samantha scolded. "We do it all the time."

Turning off the pavement onto a road covered in a red sandy clay, they drove for nearly a mile until they pulled up a white sand driveway and drove up to a one-story home tucked in the pines. They were greeted by two barking dogs, a yellow Lab and a Basset hound, and Rover barked eagerly in response.

"Don't let him out until I get a leash on him. If they start running, they might not come back until morning," Fred ordered. "Once we get them inside the fence, we can let them go."

Attempting to restrain Rover while the other dogs bounced around him and sniffed to gain what intelligence they could about their city cousin, Fred led the pack through the gate in the chain link fence. Looking to make sure the other gates were closed, he released the straining dog who immediately tore around the yard at top speed with the other two in mock pursuit.

Robert came down the ramp from the front door and waved them over, "Come on in. The dogs told me you were here. Charlotte has everything about ready."

Fred waited until the others entered and then introduced Aron. "Dad this is Aron, Sam's fiancés."

Reacting in what perhaps is best described as slack-jawed amazement, Robert gave the approaching figure a look up and down as if he had suddenly encountered a circus clown on the town's sidewalk.

"Dad. Dad, I can explain. Aron does not usually look like this. There was an accident with his suitcase just before we left, and he had to borrow these from a neighbor, I'll tell you about it when we get inside," Fred interjected.

Aron hesitated, not knowing whether to approach or start running for his life.

"O.---K. if you say so." Fred replied with a bit of doubt in his voice. "Come on up. Dinner is almost ready."

Aron approached and extended his hand which Fred shook with a somewhat weakly voiced, "Glad to meet you."

Stepping inside the modular home Aron was greeted by a mixture of Victorian furniture and antiques in glass cases along with the hunt-country South. There were paintings on the walls, a black bear rug stretched across the dining room table, a shotgun standing beside the door and statues and figures of various breeds of dogs scattered around the living room.

"Have a seat on the couch while the women get everything ready," Fred invited as he sat down in a recliner and switched off the NASCAR race that was playing on the flat-screen TV. "The bath is over there if you need to wash up."

As he looked into the bathroom mirror Aron realized that he had broken into a sweat and not only washed his hands, but also splashed some water on his face and wiped it with a washrag. Taking off his jacket, he emerged more comfortably dressed in the Martin Luther King long sleeve shirt and prison stripes.

Food in bowls had been set out on the kitchen's center isle and a stack of paper plates on the kitchen table.

Samantha introduced Aron to Charlotte, "This is Aron, the man I am going to marry," she told her step-grandmother. "We are going to be married in Detroit, and I hope you can come to the wedding."

Charlotte, now getting a close-up view of Aron for the first time, admired his facial features, slim body and noted the height disparity between the two. In her mind they looked more like father and child than husband and wife. His outlandish looking clothes appeared more appropriate for a member of a rap group on TV rather than someone who might work at a hospital. Nonetheless, all this would wait. The food was here, her guests were here, and it was time to serve it.

"Aron this is bar-b-que, Brunswick stew, potato salad, fried cornbread, a string-bean casserole, rice, deviled eggs and you can have iced tea, water or a diet Dr. Pepper to drink. The salad bowls are on the table. Please help yourself and have a seat. Go ahead and start eating. I don't want the food to get cold," she instructed.

Sam guided Aron through dishing up the food and put rice in a plastic bowl and put the thick Brunswick stew on top of it. The more normal things he served himself, but was puzzled by the thin pancake sized bits of fried cornbread.

"What do I do with these?" he asked.

"You eat them with your fingers with or without butter or syrup," Sam replied. "They are dry and crunchy around the edges. These are often served with fried fish."

Aron was happy to let Fred, Lollie, and Samantha explain his present predicament as to how and why he was dressed as he was. Thankfully, Fred did not reveal the bathroom happenings at the Flying J.

Charlotte asked Aron what he thought of the bar-b-que and Brunswick stew.

"I have had bar-b-ques before, but they were mostly Kansas City style made from beef or Chinese or Korean cooking where the meat is done over wood coals like you do, but seasoned differently. This is the first time that I have had this style of bar-b-que with a tomato sauce," Aron responded.

"Even here in the South the bar-b-ques are different. In North Carolina they eat it with a vinegar sauce, in other parts of the state they prefer a mustard sauce and each pit-master might use a different mix of sugar, honey, whisky and mix of peppers and spices to come up with a distinctive product. Usually, it is served with iced tea, even during the winter," Fred explained.

"You'll ready. I am going to let the dogs in," Fred said as he got up and opened the sliding glass door.

In came Hera, Bonnie and Rover with tales wagging.

"You can feed them a little bar-be-que if you like. With Rover you have to be careful, but Tyler taught Hera to 'take nice' and she will take it very softly from your fingers. Don't hold it up or she will snap at it. Put it in her lips and she will take it. They even know how to eat from spoons and forks. We have well-mannered dogs," Fred asserted.

Tentatively, Aron held a piece of bar-b-que out for Hera who stepped up took it in her mouth and gently pulled it from his fingers. Doing the same for Rover, the dog opened its mouth and lunged. In self-defense Aron flipped the meat into the dog's mouth and quickly withdrew his hand.

The meal concluded with coffee and red-velvet cake with the three couples sitting around the living room.

"What's the two-story house next door?" Aron asked.

"That is Whitehall," Fred replied. "That is the last of the family homes. It was built in the 1790s added to in the 1850s and Tyler fairly well rebuilt it in the 1980s. The yard was all grown up and the house was ready to fall in when he took it on. He and his wife lived there and both of them worked for the same kaolin company in Washington County."

"Kaolin?" Aron asked, "I never heard of it."

"It's also known as China clay. It is used to make porcelain china, to coat glossy papers and as a filler in plastics, rubbers and other materials. It is a billion-dollar business in Georgia. They mine a lot of it and there are several large processing plants in Sandersville. Sometimes when you come, I'll ask Tyler to give you a tour."

"Can we see the house?" Aron inquired.

"You and Sam can walk over, but take Hera with you. She is snake wise and will warn you of them."

"You have snakes?" Aron queried with a note of fear in his voice.

"Yes, we have lots of snakes, but only three varieties of rattlesnakes to be scared of. We have the eastern diamondback, cane break rattler and copperhead. I kill rattlers, but don't bother the rest of them. They help keep the wood rats and mice down. Just stay where you can see your feet and you will be all right. While you are gone, I'll see what I can set up about the wedding."

"It's nice now. I thought it would be hotter," Aron remarked as they walked towards the white plantation house.

"Spring is fairly comfortable before the bugs come out and the heat builds up. In mid-Summer there can be days of 90-degree weather with mosquitoes and gnats. This house was built on a hill to catch any wind and also had lots of windows and tall ceilings. When Tyler redid the house he told me that he closed up a number of windows and doors to make the house easier to heat and cool."

"What's that room off the back?" Aron asked.

"That is where he has his office. That was where the old kitchen was. They built it there to keep the heat from cooking out of the house and also to reduce the danger of fires. He has his library in there with some of the old family books that his dad read as a child and reference books accumulated by his grandparents and aunt who were teachers and school administers."

"I suppose there were barns, sheds and other buildings?" Aron commented.

"Most of them were torn down years ago, and Tyler took down the rest of them. Tyler described it as a nearly self-sufficient operation where the plantation milled their own lumber and ground their own grain at the mill, raised much of their own beef, hogs and chickens, produced their own vegetables and grew cotton. The only things they bought to support the 25-odd people on the plantation were manufactured goods, salt, sugar, coffee, cosmetics and things like that. Self-sufficiency was considered a virtue, and each plantation owner prided himself with making as many of the things they used as was possible."

"They owned slaves?"

"That was the economic model of the day. It was slave labor that sawed every board, made every brick, cut each piece of stone, plastered each wall and painted each room. Their cooperation, willing or unwilling, made this plantation work. I don't know the details, and Tyler doesn't either, but they had their own houses and each had a garden that they tended for themselves. I would like to think that they were beneficent slave owners, but no one knows. It was to everyone's advantage to somehow make the system work as horrible as it might seem to us today. Those were the times and those were the conditions that they lived in, and they did the best they could through good times and bad."

"This was like a Gone-With-The-Wind thing?" Aron inquired.

"Not really. Just like we have corporations today that we work for, slavery was the corporate model of its day. The plantation owner and his family were responsible not only for the plantation's economic survival but also for the well-being of all its slaves who represented a significant cash investment. It was always to his best interest to keep his slaves happy and productive. The downside was that they had absolutely no privacy. Every drop of water and each stick of wood put into a fireplace was brought there by a slave. Each horse, mule and cow had to be fed and watered each day. Every meal had to be preplanned ahead of time. Almost nothing we think of as spontaneous could happen without involving three or more people. Have no doubts about it. Those who ran these plantations really worked every day. Only the very rich who could hire overseers could live the lives that you see in the movies. The average plantation owner had to work to keep his business going. They were vested in every way in its and their family's survival. Unsaid, but always feared was that the very hands that served you supper that evening could cut your family's throats during the night. Now let's go see them."

Chapter 3

Getting To Know You

"Here are Joseph Henry and his wife Susan Elizabeth." Samantha said. "There was a brick vault built over each of their graves and his collapsed breaking the top of his cast-iron coffin. Beneath it was a glass coffin that held his body. Like most men in the South, he served during the Civil War. Because he had a bad leg, he was an officer in the Quartermaster Corps and was responsible for bringing three wagons of supplies into Sandersville after Sherman burned the town. Susan Elizabeth died shortly before the war and is buried beside him. Uncle Tyler had this concrete poured over them to protect their graves.

"He said that when last he saw him, Joseph Henry was doing well. He was somewhat gaunt, but in his uniform, although it was not fancy – just a plain grey uniform with brass buttons. There is a picture of him inside the house. When Tyler is back, I am sure he will be happy to introduce you. I think that Susan Elizabeth or one of their daughters is still around. Everyone who has lived in that house claims it is haunted. Thresa said that she saw her one night standing over Tyler's bed when he was sick. She was wearing her hair tied up in a bun on top of her head and wearing a hooped skirt with an apron. She was sweating as if she had been working and looked concerned about Tyler. Whoever she or they are, they are apparently very protective of this house."

"Wow. That is quite a story," Aron said.

"Not only that, Samantha continued, the Confederate gold that Jefferson Davis took from Richmond was reputed to have been buried on this property. They passed on the road below the house. In the 1920s a large hole was dug beside the road in the middle of the night, supposedly to recover the lost treasury of the Confederacy. At any rate, we did not get it, and we don't have it."

"Double wow. I can see why your uncle was not ready to let the place fall in." Aron agreed. "Where does that trail go?"

"That leads to one of the food plots that he plants for deer and turkeys. It's not far. We can walk it if you like," Samantha whistled, and Hera came running eager to see what new things and creatures she could find.

"These are not like our northern woods at all," Aron remarked.

"There are many more types of trees, vines and weeds. This side of the trail was recently burned on purpose and is overgrown with all kinds of vegetation. The wildlife really love it," Samantha replied. "On the other side there are more mature trees like these oaks that may be 300 years old. These were growing along the field edges when this land was last farmed during the 1950s. After that the property was planted in pines. Tyler said that these worn-out soils could just not grow enough to be profitably farmed. Twenty years later some were cut and the money used to send him and his sister to college."

Samantha picked off a honeysuckle blossom, put it between her lips and sucked the sweet nectar from it. Then she gave another to Aron.

"This is sweet. Just like you," Aron responded.

Entering the food plot, they walked through emerging grains and flowering clover. Off to one edge wild daisies, goldenrod and dogwood added to the pallet and Aron said, "I want to take a picture of you."

Positioning his phone on a clump of earth, he and Samantha sat down in the clover and held hands while the camera recorded their activities. They kissed, and very soon one thing led to another, and they had an impassioned love making on their clothes with their bodies warmed by the sun.

"We better get back before they come looking for us," Samantha remarked. As she brushed her hands across Aron's bare back she turned him around and took a closer look. "Uck. Ticks. You are covered with them." Looking down at her own legs, she said "I am too. We have to get these things off of us and the clothes washed. Ticks can carry Lyme disease and all sorts of other stuff. See, here is a big one" she said as she plucked a Texas star tick off Aron's chest.

Hurriedly dressing, they rushed back to the house where Samantha announced to Fred and Lollie, "We got into some really bad ticks, and we have them all over us. I don't even want to come in the house wearing these clothes."

"It will be all right if we can get them off you before they bite," Fred said. "Take your clothes off outside and put them into these plastic bags. I'll bring towels, and you can go straight to the shower while mom washes your clothes. Get started in a hurry because with the shower and washer running you will run out of hot water."

Handing the towels and trash bags through the door the clothes were collected and the couple sent off to the tub-shower in the master bath. They got into the shower, and proceed to scrub each other as completely as possible. Despite their efforts, their skin was still studded with pepper dots where seed ticks were digging in."

Stepping out, Samantha announced, "We are going to need some help."

Lollie asked. "Charlotte, do you have some sheets to put on the beds? I'll work on Sam in one bed and Fred can do Aron on another. Then we can wash everything."

Assuming that Hera was equally infested, Fred washed the dog on the porch. By the time the towel-clad pair emerged from the shower, Hera was shaking herself dry.

"I don't have any fingernails to speak of, and I am going to need some tweezers and alcohol." Robert said. Putting on his trifocals he walked up to Aron and pinched up a fold of skin on his shoulder. Moving his glasses up and down on his face he asked Charlotte. "Bring me that magnifying glass. These seed ticks are so small I can't see how to grab them."

Soon the pair were naked on beds in separate bedrooms while being worked on by Fred and Robert and Lollie and Charlotte with the washing machine chugging away in the background. Hera, seemingly rejuvenated, was engaged in chasing and keep-away games with the other dogs.

"Samantha, there is no need to tell us what you were doing out there, I can guess," Lollie said. "Just remember you do not make love in the woods again; however romantic the situation might be."

"We really didn't plan to. It just, well happened. Hera was off in the woods, and we just did it," Samantha explained. "I've been on birth control ever since we met, so everything is all right, except for these bugs, I mean."

"I'll have to talk to Hera about being slack in her human supervision duties," Lollie said. "Bugs we have and bugs you got. You just don't need to take any back to Detroit, although we do have plenty to spare."

A series of chimes from the washing machine indicated that the cycle was over. Charlotte went into the utility room and put the couple's clothes into the dryer. When she returned, she said, "I am going to have to press those pants and that shirt Aron was wearing. I hope they don't shrink. As soon as they are done, you are going to have to start back to Atlanta to get to your dad's at any reasonable time."

Aron was not feeling any more comfortable with the de-ticking process than Samantha.

Sitting up in bed on the sheet he was picking ticks off his genitalia while Fred was working on his back and Robert had started on his left foot and was working up his leg.

"Aron I 'm sorry if I pinch you," Robert said. "Between the glasses and the magnifying glass, I can't hardly see the ticks well enough to pull them out. I am getting the big ones well enough, but someone with better eyes than mine is going to have to get the tiny ones. Fred, are you having any better luck?"

"It is slow going, but I am getting them out," Fred replied. "I think Samantha needs to call Tom and let him know what's happening. I always had doubts about his hot-tub thing, but maybe it can be put to good use. Put some Clorox in it to kill them and they should be easier to get out. Their hair might be a couple of shades lighter for a month or so, but that might work."

"Are they in my hair too?" Aron asked.

"I imagine," Robert replied. "You can try Hera's flea and tick shampoo on your head, wash your hair and comb it out, and that might get them. Maybe then Rover might accept you as a member of the pack. I have a good set of electric dog clippers. I could cut your hair, and then do the doggie shampoo."

"Get the clippers," Aron said. "I sometimes wear my hair short anyway. Bald would make little difference."

Robert returned with clippers, a pair of scissors and a comb. Kneeling on the bed with his knees against Aron's hips he asked one last time. "Are you sure you want me to do this?"

"Yes." Aron replied firmly. "Cut it all off."

"I told Sam to call her dad and tell him about your predicament," Robert informed.

"I have the finest set of teeth on the clippers. This will not shave you, but it will give you a close bristle cut. I've washed the teeth with alcohol, so they are clean." With a steady push he started at the nape of the neck and moved forward until he reached the forehead, and then started another cut, much as if he was mowing a lawn. Soon the bedsheet was littered with one-inch strands of Aron's black hair.

"How did that feel?"

"You are doing fine. I have had barbers do worse," Aron responded.

About a half hour later, there was a knock on the door to give Aron a chance to cover up, and Charlotte came in with Aron's clean, pressed and folded clothes. She looked at Aron's fresh haircut and surmised what had occurred.

"Robert that looks good. If you did that well with the dog clippers you don't need to go to the barber shop anymore," she commented. "Sam called her dad and told him what to expect. He said that he would be ready when you came."

Least it be forgotten in their rush to leave, Charlotte told Samantha and Aron that she and Robert would go to the wedding. When the date was set, they could let them know and Fred could get their tickets.

"I still feel like things are still crawling on me," Aron told Fred as they were driving up.

"They probably are. When we get to my house you can load up your rental and go to Tom's. It is sometimes awkward between me and him, but for Sam and Beth's sake we try to keep on speaking terms. All in all, it has worked out fairly well because I met Lollie after they were divorced."

"What happened between me and Tom is that we found that we just could not get along," Lollie interjected. "We were fighting all the time. We decided to separate under a joint custody arrangement, rather than continue a marriage that neither of us was happy with. He and Susan are a much better match that we were, and I am happy for them both."

Arriving back in Covington, Charlotte hurriedly packed Aron's bag and put a strap around it because the zipper had burst. Aron recovered his billfold, and he helped Samantha get her things together. Rover, delighted to be home, was busily checking the back yard. When Sam and Aron went to leave he went up to Sam for a parting pet and then sat in front of Aron.

"He wants you to pet him," Samantha said.

"He won't bite my hand off," Aron replied hesitantly.

"No. He knows that you are part of the family pack now," Samantha said reassuringly.

Aron held his hand in front of Rover's nose, and he licked it and allowed him to reach behind his head and scratch the back of his ears, like Samantha had when they arrived.

"My dad would never let us have dogs," Aron remarked. "He said that they were filthy, dangerous animals; and he would never let us bring one into the house."

"See. Just like I said. It just takes him a little time for him to get use to people."

Now loaded in their rental car, Samantha drove and Aron prepared himself for yet another adventure in family living. He remarked to Samantha "It's gotta get better. It can't get any worse."

"Margaret, Elizabeth, come down here," Tom called to his daughters.

"What's up Dad?" Margaret asked because it was too early for supper, and both of them had been doing their homework.

"What do you two have for tomorrow?" Tom inquired.

"I have some chapters to read for a couple of courses," Margaret replied.

"I have three or four more algebra problems to do and some reading," Elizabeth responded.

"Good. Then you can help Sam and Aron. Down at Fred's parents they went into the woods and came back covered in ticks. They got bunches of them off down there, but they are going to need our help. I'm going to make up a chlorine solution in the hot tub and after they soak in for a while you can help me and your mom pick them off. Go get your tweezers, some alcohol, cotton swabs and a magnifying glass and come back down," Tom ordered.

At a loss for words for once, the girls went back upstairs and gathered their best guesses at adequate tick-picking materials and returned.

When the back doorbell rang, Samantha and Aron were immediately taken to the bathroom where they undressed and got into the hot tub. Susan immediately took their clothes and put them in the wash.

"I made up a weak chlorine solution for the hot tub," Tom informed the pair. "This is about the same as they put in swimming pools. It should kill the ticks and maybe the heat will cause them to release. Aron where did you get that haircut? You look like a Marine," Tom asked.

"Robert did it with his dog clippers," Aron replied.

"I've asked Margaret and Elizabeth to help," Tom started with a bit of hesitancy in his voice. "I need to talk to you about that. Margaret has already had sex and is on the pill, but Elizabeth hasn't and isn't. I know because Margaret and her boyfriend did it in this house. I want my daughters to know about sex and be safe about it. I had rather they have these initial encounters here rather than in a back alley or parking lot with one of them coming home pregnant before they even had a chance to start their lives. This introduction to sex thing is what Susan and I worry about most. I can't say that any parent ever got it completely right, and I don't claim to either. When you have your own family, you will know what I am talking about. Sam, Susan and I are going to be here with you. If the girls have questions, particularly Elizabeth because she is full of them, please answer them as best you can; but if they get too personal say so. They don't need to know the details of your sex lives, but if they ask reasonable questions, please respond to them as best you can."

"This is getting very close to pornographic acts and child abuse, but if this is what you want, I will go along," Aron reluctantly agreed.

"Dammit, it did get worse," he thought.

Susan came in with some blankets and spread them and two pillows on the floor. Handing towels to them both, she said that they should dry off and lie down when they were ready. Aron dried himself and wrapped one of the towels tightly around his waist and lay down.

As Susan left with the wet towels Tom, Margaret and Elizabeth came in.

"You are a big one aren't you?" Elizabeth remarked as she saw the two laying side by side.

"Sam why did you pick a guy that was so big while you are so small?" she questioned with a sound of concern in her voice. "You are going to have some enormous kids."

"I put up with being the smallest kid in the school, in my classes in college and so on all of my life. I was determined that whoever I married was going to be bigger than me so my children would not have to go through the same thing. Besides, I just happen to love him," she concluded as she reached over and tickled him on the ribs.

"Now get to work you two," Tom ordered. "You know what to do."

Beginning on Aron's left foot where Robert had left off, Margaret quickly started pulling off the ticks with first the tweezers and then with her fingernails. Looking at one she had extracted she said, "Dad, that chlorine worked. They don't seem to be moving."

Sam was getting the same treatment from Margaret who was more interested in learning about the coming Detroit wedding. "Where is the wedding going to be?" she asked.

"I don't know yet," Samantha responded. "Aron's father and mother are planning that. We have agreed on a Jewish ceremony, but where it and the reception will be, I just don't know. I hope that you and Elizabeth can come. Aron has scads of relatives, but we have comparatively few. When I come down for Elizabeth's graduation, things should be fairly well planned out."

"Elizabeth, how are you doing over there?" Tom asked.

"I am up to the towel Dad. What am I supposed to do?"

"Aron, I am afraid that you are going to have to take it off," Tom replied.

"Lay still. I'll do it," Elizabeth replied as she gently untied the knot and laid the towel on both sides of his hips. "You are big. How do you and Sam…"

Aron held up his hand in a "stop right there" motion and Elizabeth reframed her question.

"I hear that you work for a conception and fertility clinic. I have a kid in my school who has two men for parents; and he says that he is their child. How did that happen?" she asked with a tone of genuine interest.

Considering that this was an answerable question, Aron replied, "We assist gay couples of both sexes to have children. For the men we mix their sperm and implant it into a foster mother's womb. She carries the child and when it is born gives it up to the male parents when it is old enough that they can care for it. That retention time may be months, depending on the child and the circumstances of the birth. It is easier for two women. A sperm donor or donors are found, and one of the two women impregnated, and the birth and delivery go as you might expect."

"Who are the sperm donors?" Elizabeth continued.

"Often, they are students who are in the medical profession. It is assumed that anyone who makes it into med school is reasonably intelligent and they pick up a few extra bucks each time their sperm is used. At one stage the donors were kept anomalous, but with the availability of genetic testing, some people have found out that they have up to 40 half-brothers and sisters scattered around the country."

"That would come as quite a shock," Susan responded. "I could imagine that would be quite a family reunion with everyone looking somewhat alike and being about the same age. It would be like meeting yourself as you came around a corner."

"Were you a sperm donor?" Elizabeth probed.

"No. When I went to work for the University Hospital they did extensive tests. One of the things that they found out is that I have a defective chromosome that if combined with the same defective from a woman's chromosome the child has a high probability of serious and often fatal diseases."

"And why didn't you tell me before we ever had sex," Samantha exploded. "It wasn't until we started thinking about having children that you told me. Don't you think that I would have been interested in knowing that?"

"There would have been no trouble if you don't also have a matching bad chromosome. We'll have things checked out as soon as we get back to Detroit," Aron interjected. "This is not a common genetic defect and chances were that you did not have it."

"That is something I would have wanted to know," Samantha asserted.

"I never knew they looked like that," Elizabeth said pointing to Aron's wrinkled scrod. "All I had seen on those statues were smooth, and I thought Dad's was wrinkled jest because he was old."

Tom fielded this one. "Those sculptors made them smooth because they were easier to carve. When a guy has an erection, it draws up tightly against the body to help project the sperm and the fluids that carry it out of the penis. Sorry, no demonstration. Right Aron."

"Not if I can help it."

"I am done on this side now turn over please," Margaret requested. As she worked on his buttocks she paused and said, "Dad would you please get that one," pointing to a tick firmly attached to the spot where the sun never shines.

Mercifully done with the tick picking and smelling of rubbing alcohol and chlorine, Samantha and Aron prepared to go upstairs to bed.

"Would you like a drink?" Tom asked the pair. "Maybe a hot rum totty."

"Like we have for Christmas?" Samantha asked.

"The same, I have the Hudson Bay Rum, spices and butter. Come upstairs and I will get them ready. After today, I think you both need a little something."

For once, and perhaps the first time, Aron was in wholehearted agreement with his host.

Chapter 4

Medical Issues

Dr. Bagwar Austongone Genetics was the name on the door of the waiting room when Aron and Samantha went in to receive the results of Samantha's blood test. Both looked nervously at the certificates on the wall from Asian universities, and pictures of water buffalo plowing rice patties and temples with exotic statuary.

"Come in. The doctor will see you now," the receptionist announced. They arose and went into the examination room. In the room was a human DNA model and a skeleton suspended from a steel hanger in the corner.

"I wonder if that was one of his patients?"

Samantha asked to break the tension.

"Those are fairly cheap in Asia." Aron explained. "Most of them come from China, and he has probably had it since he was a student. Many doctors become attached to them as non-verbal colleges. This is a strange doctor-tool relationship. It is considered an honor for one physician to pass his skeleton on to a colleague and particularly to a student."

After a knock on the door, a balding browned skinned man entered carrying a folder containing a sheaf of computer printouts and graphs.

"Hello, I am Dr. Austongone. My specialty is human genetics and in particular genetic-transmitted diseases and child defects. I have had a chance to examine both of your blood samples and have run them through the process twice just to make sure of the results. I am afraid that I have some bad news."

"What? What?" Samantha demanded with a note of fear in her voice.

"As you know this man who I understand is to be your husband has a chromosome defect that carries with it the possibility of kidney failures in infants and young children. This is an uncommon condition and occurs perhaps once in several hundred individuals in the U.S. If this defect is reinforced by a sex-partner having the same type of defect, the children from such a marriage often die at a young age.

"This condition is known as Haemochromatosis which results in an abnormally high concentration of iron in the blood. Blood iron is necessary for your health and for that of the developing fetus. There are two defects, and one is not as serious as the other. Unfortunately, both of you carry damage in the C282Y part of the chromosome which means that your children will have a higher probability of having severe liver damage and associated bone and organ diseases which can result in still births or shorten their lives to a matter of a few years.

"This disease is more common in some western European countries where it has been intensively studied. In Scotland, for example, one in every 114 children will be born with the disease. With the more diverse genetic pool in the U.S., the odds are not that bad, but I can't tell you exactly what they are. A lot of information about Haemochromatosis has been published in the U.K. and is available online, but not so much in the U.S. The average person may have heard of Sycle-cell defects of blood cells in black populations, but not Haemochromatosis."

"So this not a particularly Jewish thing?" Aron questioned.

"No. Many non-Jews also carry the defect," Austongone replied.

"Is carrying this defect going to shorten our lives?" Samantha questioned.

"In adults, typical symptoms are a bronzing or graying of the skin, loss of body hair, fatigue and loss of the sex drive in men among many others. Since you are carriers and only have one copy of the defect, neither of you have these symptoms and you will not develop them. You should live normal lives. In cases where adults are impacted, periodically drawing blood and letting the body replace it with newly generated blood cells is a common treatment. This is one case where the old medical practice of bleeding a patient actually had some benefit.

"Samantha, there are ways that you can have healthy children. Not all of the chromosomes in eggs and sperm carry the defect. There are options. You can use sperm from a sperm donor, or we can select sperm and eggs that don't carry this defect, do in-vitro fertilization and implant the fertilized egg. This is a relatively new technique, but our doctors have had a high rate of success with it.

"Aron, if you chose to use a sperm donor, a brother or cousin who are not carries of the defect, might be good candidates. Otherwise, you may select from donors who best match your physical profile."

Samantha said to Aron, "What are we going to do? I want to have your child – not somebody else's."

"You don't have to make a decision right now," Dr. Austongone said. "You are fortunate that your husband is a hospital employee and the costs of either procedure will be paid for by his insurance. Using his sperm is more expensive, but the real decision is about the risks that you are willing to take. You need to talk to the surgical implantation team about those, and I will set you up with an appointment."

"If we are successful and have a child using my sperm, will he carry the same defect?" Aron questioned.

"Although we are getting somewhat closer to absolute predictability we are not there yet. The probability of your child passing this defect onto his children is small based on the results that we have achieved in similar cases. We test before fertilization, during fetal growth and after delivery. If all these tests are negative, the possibility of his developing the same defect is no greater than anyone else in the population."

"I have another concern," Samantha said. "What is this about COVID-19. It's all over the news. Is it going to get to the U.S. and be like the Spanish Flu after World War I? That killed millions."

"I have seen cholera and malaria rage through India, and those were terrible experiences, "Austongone replied. "I don't know any more about COVID than you do. Look out for what the Centers for Disease Control has to say about it and follow whatever recommendations that they have. In the meantime, make sure you stay on birth control and as an extra safety measure use condoms when you have sex. All of this can be safely handled one way or another, but you need to be careful."

By the time Samantha and Aron had their appointment with the implantation team a week later, COVID-19 was raging in New York City and death rates were rapidly rising. Even in Detroit hospitals beds were starting to fill with COVID patients and there was talk of canceling all non-essential medical procedures to free staff and beds for the growing numbers of deathly ill people.

Their wait for Dr. Mayfield was in an even more antiseptic setting with steel chairs and blank, white-painted walls.

"Samantha, this is sort of scary. Are you sure you want to go through this?" Aron gently inquired. "After all we aren't even married yet?"

"We have talked through this before. With this lockdown business, and everyone being shut up at home and hospitals filling up, they need to do this today. They are just going to take the eggs and sperm and put them in cold storage until after we are married and then implant them. I know this sounds like some Dr. Frankenstein type of thing, but we need to do this while they have the people and facilities available."

Dr. Mayfield came into the room. He was wearing a mask and surgical scrubs. "Samantha, Aron you have had some preparation to do for this procedure, and I need to ask you some questions that you may find embarrassing.

"Aron, you first. Did you masturbate two days ago? This is necessary so that you can generate the maximum number of active, healthy sperm."

"Ur. Yes. Sam helped me. We did it. Because it was beaten into me that this was such a terrible, sinful thing to do, I have trouble doing it by myself."

"We can extract sperm from you if I need to, but it is easier if you ejaculate into this sterile cup. Catch all of the semen in it and immediately close it up. Wash yourself and your penis thoroughly before you do. Or I will have a nurse do it for you. It will not be necessary to shave you unless I have to take the sperm with a needle.

"Nurse Elrod, would you bring a pan some surgical soap and a washrag?" Mayfield asked.

A large black women entered the office carrying a stainless-steel pan with a rag and a small plastic bottle of green surgical soap.

"Aron you can go into the bathroom and get started while I talk to Samantha," the Doctor ordered.

"First we are going to do an ultrasound and a blood test to make sure the eggs are ready to extract. If everything is good, I will go through your vagina with an probe that has a needle attached that will allow me to see inside of each ovary and suck out the eggs that the drugs stimulated. Sometimes there is some pain and bleeding that follows. We are going to keep you here for about half an hour to make sure everything is all right and then let you go home. You should not drive or do anything strenuous today. By tomorrow you can return to your normal activities. These preliminary tests can take about half an hour, and then we can proceed. Do you think that you want anesthetics?" he questioned. "Insertion of the probe will be a little uncomfortable, and there may be a small amount of pain."

"Will there be much bleeding?" Samantha asked. "I did not bring anything."

"Not usually. If you continue to bleed after you get home or if you start feeling very weak come back to the hospital immediately. Most often no one has any problems unless they have some preexisting problems which prevent their blood from clotting."

After Samantha was taken in for prep, Mayfield walked over to the bathroom door and knocked. "How are you doing in there?" he questioned.

"Not good. This is all so strange," came the reply through the door. Inside Aron sat on a stainless-steel commode surrounded by stainless steel walls with a drain in the middle of the floor. The bathroom was so tiny that he had to squeeze between the wash basin and the wall to get to the toilet. Sitting down with his pants down on his shoes, his member hung like a deflated balloon and stubbornly refused to respond to his manual stimulation.

"I will get you some magazines. Maybe they will help," he offered.

Returning with a stack of soiled, much thumbed magazines that looked like they had been recovered from a bus station's trash can, he opened the door slightly and pushed them in. The top one was titled, "Big Momma Sex."

"Thanks," Aron said as he squirmed past the sink, got his feet caught in his own trousers and he crashed heavily against the door hitting his head.

"You all right," Mayfield asked as he opened the door and found the semi-undressed figure sprawled across the floor. Helping him up, Aron stood unsteadily on his feet with one hand holding his head and the other reaching for his pants, and he nearly fell over again.

"Nurse Elrod, come here," the doctor shouted. When she arrived, she grasped Aron's left arm while Dr. Mayfield had the right and reached down to lift the pants from his ankles. The pair half drug Aron into another exam room where he was laid on an exam table which had attached stirrups suitable for his usual female patients.

"Are you all right? You had a hard knock on the head."

"I think so," Aron replied. realizing his state of undress, he made a move to sit up, but was restrained by the Doctor's hand on his chest.

"You lay there and be still. I am going to take the sperm from testes with a needle.

"Get him undressed, cleaned up and shaved," he told Nurse Elrod. "After I have finished with Samantha, I will come back and take the sperm."

"I can't imagine a big man like you having any problems," the nurse stated as she pulled off his shoes, socks, trousers and underwear. Then she mixed some shaving soap in a cup and stropped a straight razor on a leather strap. Laying a warm wet towel over his midsection she stepped up to look him in the eyes.

"For coarse hair like you have down there nothing is as effective as this old-fashioned straight razor. I do this all the time. Just lay back and be still," she said in soothing tones. "Your girlfriend is going to like it when I finish."

Adding to his aching head, he now felt his face flush as if it were scalded and realized what he must look like laying there with his large feet pushing on the stirrups while the nurse shaved him with that extra-sharp razor.

"Thank God. There was no one around to take a picture," he thought.

"Samantha, are you ready?" Mayfield asked.

"Yes. Go ahead," she replied.

Samantha felt the tools go inside her and watched as the doctor looked at a screen to see the process. She could not see his hand movements but could feel something as he shifted his fingers slightly to move from one ovary to another.

"I am almost done," Mayfield told his patient. "We will select several healthy eggs and preserve them until it is time for them to be fertilized. I am going to take another look to make sure there is no bleeding and after you recover a bit, you can dress and leave. You have plenty of time. I will send my nurse to help. Just lie there and rest for the moment."

Returning to Aron's room he sent Nurse Elrod to assist Samantha and then searched through two drawers to find a fresh syringe.

"I am going to push this through your scrotum into a testicle and draw out some fluid that contains your sperm. Then I will select the most active for further testing and it is from those that some of Samantha's eggs will be fertilized. We will keep these so that in later years you would like to have more children we will already have the fertilized eggs to implant."

When Aron looked up and saw the needle, he fainted, and his head fell back on the table.

Putting down the needle, Dr. Mayfield moved Aron's head to make sure he was comfortable. He then swabbed the exterior of the scrotum with alcohol, prepared a new needle and then pinched a testicle firmly in his fingers. Feeling the needle meet resistance, he pulled back on the plunger and extracted the sperm.

Aron awoke to the biting smell of ammonia irritating his nostrils which provoked an involuntary sneeze that shook his entire body.

"Be still. You have had quite a day," Mayfield said. "I have extracted the sperm, and they look good. Samantha's procedure went well too.

Nurse Elrod will help you get dressed. I need to make sure that you are both all right before you leave. How is your head, balance and eyesight," he concluded.

"They are fine, I think I am all right," Aron responded. "I'll know better when I try to walk."

"What are those bumps all over your and Samantha's skin. They look like insect bites. Do you have bedbugs at home?" the Doctor asked.

"No. Those are tick bites. We got into a bunch of them in Georgia, but we got them off of us. Neither of us caught anything from them," Aron said.

"You are lucky. As a couple, you don't need any more health issues right now." Mayfield concluded.

Aron was driving back to the apartment when Samantha asked, "How did things go with you?

"Not especially well. I could not do it in that bathroom. When Dr. Mayfield was handing me some porn magazines I fell and banged my head against the door. They took me to an exam room and the Doctor took the sperm from a testicle with a needle. He said that the sperm looked fine."

"You don't have any trouble doing that with me," Samantha observed rubbing him on the thigh.

"No I don't. In fact he is telling me we could have sex right now, but the doctor said to wait for three days until we both were very sure we were over the procedures.

Chapter 5

Working At Home

"How long is this COVID lockdown supposed to last," Samantha asked Aron.

"I don't have any idea. The hospital said that since my job does not require patient contact, I should plan on working from home until this epidemic is under control. They are working on a vaccine, but that is months away. They said that a tech person would transfer my computer, relocate my phone and hook everything up.

"Where in the world are we going to put your stuff?" Samantha puzzled. "I already have my computer, printer and desk in the extra bedroom, and we can't both be in there because we would be talking over each other all the time."

"There is that hall closet," Aron suggested. "If I put a chair inside the closet and a small desk in the hall and a box or something to put my phones on, maybe I could work from there, but it is awful hot in there and there is almost no air. That would be like working in a coffin."

"The only real table that we have is in the kitchen," Samantha offered. "Maybe you could use half of that. Your chair would be sticking out in the living room. It doesn't look like we are going to be having much company, and if we did, we could move your computer off the table for a day or so."

"If we move the bed against the wall and then put the dresser beside it, I could put a slim computer table against the other wall and push the chair up against it when we went to bed," Aron supposed. "That way I could close the door and reduce the noise. There are already electrical and phone plugs in the walls."

"So you are going to lay in there and sleep through the epidemic?" Samantha kidded.

"I wish everyone could hibernate until the danger has passed, and being locked-down is about as close as Dr. Fauchi, the CDC and city health officials can think of," Aron postulated.

"I guess the bedroom is the best place for you," Samantha agreed. "I like you in there anyway." With that statement she walked over, took his head gently in her hands and kissed him.

A ring from their call box indicated that someone was at the outside door of the apartment building. When Aron pressed the button a voice replied, "Tech support. We have your computer and phone system to install."

"Come on up we are on the third floor." Aron said.

"We have a number of boxes. Do you have an elevator?" the voice replied.

"I am sorry we don't. I'll come down and help," Aron replied.

For once Aron was glad that Samantha had gone shopping to pick up toilet items and food for a couple of weeks. They had never done much cooking, but mostly ate out when Samantha was in town.

Reaching the bottom of the stairwell, Aron saw Ju Kim and Stan Lee struggling to provoke a long, tall, thin box around the corner of the first stairwell landing.

"This is your computer table," Ju said. "Grab the end and go up and Stan and I will work on the back. I'll cut some holes so you have something to hang onto."

Ju took a box-cutter and cut some holes on either side of the box just above the lettering that proclaimed the contents, "The Village People in Concert" computer table.

"What's this all about?" Aron questioned. He was puzzled about the object he was about to move into his and Samantha's home.

"Everybody is setting up to work at home and buying up computer tables and anything about setting up a home office." Ju replied. "We can't hardly find anything. We have a Mickey Mouse desk on the truck if you had rather have it. That one is kid-sized, and you look like you need the full-sized version."

"I guess so," Aron replied.

Unpacking and setting up the table, Aron saw that beside the flamboyantly costumed band pictured on the unit's table, back and shelves, it was already wired with plug-ins for two computers and telephone connections. That unit had to be completely assembled and the box and packing pieces removed to allow room for the men to bring in the computer and phone.

Next came his tower unit from work. It was set up and the monitor screen connected.

Stan observed, "This is an old Compaq with a number of out-of-date programs. This needs to be upgraded to Windows 10 and the new hospital software, but I doubt that it has enough memory. We are going to put in a new unit that can take the upgrades and let you remove any files and programs that you no longer need from the old computer and transfer your critical files to the new one."

"Oh, Crap." Aron thought. There was nothing he liked less than fiddle with programs that he did not know and go through the time consuming and often flustering installation programs.

"Here are your manuals," Ju said. "They are very straight forward."

Looking at the plastic wrapped manuals he saw that the top one was written in some Oriental language that he did not recognize.

"What's this?" he asked.

"O. That's the Korean version. There is also a Chinese one and others in French, German and English; although those in the European languages might not be so detailed," Ju concluded.

"Fantastic," Aron remarked ironically.

With another load of boxes and packing hauled away, the workstation was starting to take shape. The computers were set up on the table, his new cell phone unwrapped and a plug-in phone installed. So far as Aron could tell the jungle of wires were all connected.

"Now the smoke test," Stan stated. "I am going to do the start-ups one by one. This desk has a built-in surge protector, so you are O.K. there."

Stan depressed the computer table's main switch and a thunderous explosion of sound erupted as the opening lines of "Y.M.C.A" blasted from the unit accompanied by flashing lights from the painted figures.

Aron covered his ears and Stan, taking the hint, adjusted the volume by twisting a knob on the back of the desk. "There is an app that will allow you to control this power station from your cell phone," Stan said.

"Can I turn that stuff off altogether?" Aron inquired with a note of desperation in his voice. He could imagine what Samantha might think.

"No." Stan responded. "Each time you activate the system you will have the sound and light show. It only lasts a minute or so. You'll get accustomed to it."

Next Stan activated Aron's old computer which came up like a wheezing donkey. "Wow." Stand said, "This thing sounds like it is about to die. Your hard drive is dragging, and it is slow, slow, slow. You need to back up your critical files on the new computer and save them in the cloud. Once it quits, we can maybe retrieve some data, but no guarantees if the hard drive is damaged."

Once Aron's old computer came to an operational state, Stan started up the new unit. Flashes of neon-colored lights emanated from the drives that were visible through the glass sides of the tower and the keyboard turned on showing illuminated green letters. The screen came up with the programs listed in both English and Chinese characters.

"There are no CD ports on this computer," Stand informed the dazzled Aron. "You can transfer data through the cloud, with thumb drives or with direct connections which I do not recommend. The Hospital does not want to take the chance of any viruses being transferred from your old computer to the new system.

"Now let's check your phones. I will call your hospital number and it should ring here," Stan said as he punched in the number into his cell phone.

Another course of Y.M.C.A emanated from the new phone as it vibrated and flashed. When Aron looked at Stan, he responded, "I thought you would like the same theme on your phone. You can reprogram it to another ring tone if you wish. Do you have any questions?"

"Yes, a million of them," Aron thought, but he was too mentally exhausted to think of anything right then. "No. Thanks guys. I appreciate you doing this. If I have questions when I get into the system, I will call."

Gathering their tools, the pair left. Aron sprawled on the bed thinking he could grab a nap before Samantha returned. All too soon the apartment buzzer rang informing him that Sam had arrived with the groceries, and she needed help bringing them up.

"Next time you are going with me," Sam announced as soon as she saw Aron coming down the steps. "It was a madhouse out there. They would not allow but 75 people into the store at the time. When I went in there was nothing at all on some sections of shelves and as soon as the stocking people brought in a pallet of something they were swarmed, and people grabbed whatever it was they were putting out. People were grabbing toilet paper, bottled water and practically any meat and vegetables that were in cans. You would think there was a war going on. I got what I could, and I bought some things from the Specialty Foods Sections. I did get the basics, some fresh meat, fish, salad vegetables, canned stuff, oatmeal, breakfast flakes, milk, almond milk, beer and wine."

"I did not imagine that it would be anything like that bad," Aron replied. "I wonder how mom and dad are doing?"

"Things were so bad that a policeman with a gun walked me to my car and stayed there until I drove away. He said people were stealing things out of people's carts in the parking lot. You should call your parents after we get these things put up. Maybe it is not as bad where they live, but you might have to do their shopping for them."

"I've got to check my e-mails," Samantha concluded. "You put things up, and then call your folks."

Aron unpacked the grocery bags and put the food on the kitchen table. As he did, he sorted it into groups as it would be stored in their cupboards, fridge and freezer. The freezer stuff went in first which included some small inexpensive pizzas, pot pies and various frozen meals. The meats and fish were somewhat more challenging with two steaks with lots of gristle and some fat in them, a roast as big as a basketball, unsliced smoked and cured pork belly and a more or less foot-square block of salted foot-long fish.

Thankfully, there were vegetables and fruit that he recognized like onions, a sack of new potatoes, carrots, celery, Lettuce, apples and oranges.

"Sam," Aron called from the kitchen. "We don't have room for all this stuff."

"Do the best you can and put the things that won't spoil in a box under the table," Sam suggested. "Take that roast and cut it up into three or four pieces, wrap three of them and put those in the freezer. I will cook that other one tonight or you can start it now. I am in the middle of a conference call. I'll check on you in a few minutes when you are ready to put it into the oven."

Aron selected the largest cutting board that they had and after unwrapping the meat in the sink saw that about 30 percent of that piece of beef was fat. He then checked on the label and saw that it was called "Beef Brisket." He knew his mother often cooked a brisket for Passover, and he needed to call her anyway.

"Mom. This is Aron. Sam just came back from shopping and said that it was really terrible with long lines and people even stealing things out of shopping baskets. Do you need for me to get some things for you?"

Sarah hesitated, and Aron could imagine his mom making a mental inventory of her kitchen. "No. Not at the moment. Your dad and I have been watching TV and seen some of the mobs at the stores. We have plenty of food at the moment, and our grocer still delivers to his old clients. I think that we are going to be all right.

"David, it's Aron he wants to know if he needs to go shopping for us?"

Another voice was heard as David joined the conversation. "I need a new battery for my lawnmower and some 10-30 weight oil for it, but otherwise, I think that we are doing fine. There is no hurry. Next time you are near an auto-supply store you can pick those things up. Are you and Sam doing all right?"

"Sam just came back from the store, and she said that it was a mob scene out there, and she was frightened. I don't like to go shopping, but next time I am going to go with her. She had to have a police officer walk her to her car."

"One would never think that could happen in Detroit," Sarah considered.

"I am not at all surprised," David added. "Unless restrained by the law, people will do anything when they are stressed. And believe me, there are some stressed-out people who have lost their jobs, have no money coming in, have families to support, are cooped up inside all day and are facing a deadly epidemic. This is something like the horror days in Europe at the end of World War II. Except then there was hope in sight. Some have lost all hope that anything will ever get better."

"Mom, I have a question of what to do with a huge piece of meat that Sam brought home?" Aron interrupted. "It is called a brisket. I am going to cut it up into four pieces, freeze those and cook the other tonight. I know that you cooked it after you soaked it in something. It's got lots of white fat on it. What do we do with it?"

Apparently relieved to change the conversation to some more appealing subject, Sarah replied. "On what you are going to cook, cut it into pieces that are about 3/8ths-inch thick, marinate those in vinegar, salt and cloves for several hours and then boil it until it is tinder. Then you can add cabbage to it or whatever other vegetables that you have and finish seasoning it with a bit of coriander and some coarse crushed peppercorns. Once you have the meat tinder you can do almost anything with it."

"What do I do with all of that fat?" Aron questioned.

"Since you are just cooking for yourself and Sam, cut most of the fat away and freeze it. You can use that to season soups, stews and vegetables. It doesn't keep all that well, so you want to make sure you freeze what you don't use in a few days. It will go rancid on you."

"Thanks mom that is a big help," Aron replied. "This is the biggest hunk of meat that I have ever tried to cook."

With the fat removed from roast and sliced into more manageable pieces, Aron sliced the meat into thinner sections before marinating it in a flat Pyrodex dish. Just as he was finishing the sound of "Y.M.C.A." erupted from the bedroom. He dumped the last piece in the water, vinegar, salt and spice marinate and spilled some of it onto the floor as he attempted to jockey the dish onto one of the crowded refrigerator shelves.

"What is that God-awful noise," Sam demanded from her office.

"That's my new phone," Aron responded as he rushed to answer the phone which was vibrating and pulsing in red, orange and green colors on his new computer table.

"Aron Goldfarb. University Hospital," Aron said as he picked up the phone.

"This is Alonzo Boden, with the University Technical Department. "Did my two tech reps get your computers set up today?"

Relived that this was a business call instead of a couple needing reproductive aid, Aron happily answered. "Yes, they did. I haven't used it yet, but they tested all of the components before they left."

"I will be sending an evaluation form to your e-mail address," Alonzo replied. "When you have had a chance to use the system for a week, please send it back. If you don't I can put a hold on your paycheck. It is important that we know how these tech reps are doing with a lot of our administrative people starting to work off site."

"Will do," Aron replied. He was eager to terminate the conversation as quickly as possible.

Running back to the kitchen to clean up his mess, he arrived to find Samantha approaching the fridge.

"Don't," Aron shouted with such force that it stopped Samantha in her tracks.

"What the hell?" She replied. I just want a bottle of water.

"I spilled some marinate onto the floor, and I don't want you to slip on it. I will get your water for you," Aron volunteered as he stepped wide over the spill, opened the door and retrieved a bottle of lime-flavored water.

"Thank you." Samantha said perfunctorily. "See you later as she accepted the water handed to her by her spread-eagle boyfriend she reached and gently massaged his testicles.

Chapter 6

Graduation 1

AARON CAREFULLY REMOVED THE GLASS dish containing the marinated brisket from the refrigerator and placed it on an old towel he had put on the kitchen table. Getting a small platter and lining it with folds of paper towels, he carefully lifted each piece out of the marinate let it drip, dried it with the paper towels and placed it on the platter – just as he had seen Julia Childs do on TV.

After all of the meat had been removed, he put the marinate into a pot and put it back on the stove to reduce. Then attempting to do something he had seen at Le Monde Detroit he turned on the vent fan, uncorked the bottle of red wine and prepared to flame the meat. Looking under the cabinet he got out some peanut oil and put that in a cast-iron frying pan. While the oil was heating he shook the meat in a plastic bag with flour and pepper and when the oil was smoking, put the meat in the pan.

The smoking pan quickly seared the meat and he then poured in the wine and swirled it around in the pan as he lifted it off of the stove. As he moved the pan the alcohol spontaneously ignited and flames two-feet high leapt from the top of the pan accompanied by a blast of heat to his face. The apartment's fire alarm instantly went off. Now that the flames had subsided, he placed the pan on a cold eye just as Samantha burst into the room.

"What are you doing?" she shouted. "Are you trying to burn the house down?"

"Everything is under control," Aron responded. "I am just trying out some things that I saw on TV. Don't worry. It will be good when I am finished with it. You can go back to your meeting. I've got this."

Now proceeding somewhat more cautiously he put a half cup of finely chopped Spanish onions into the pan and deglazed the frying pan after which he dumped in the sliced carrots and potatoes to capture the last of the residual heat. in the meantime, a rising cloud of white froth was rising from the marinate pot which boiled out of the pot and ran down to evaporate on top of the stove where it left a burnt crust.

Turning down the pot, the water level subsided and his two-burner stove crisis had apparently passed. Stopping and reevaluating the situation, he dug out a pressure cooker and dumped the contents of the frying pan into it. Then he strained out all of the solids out of the marinate and added some of that liquid to the pot and then water to cover the vegetables.

"Maybe 30-minutes more," he thought.

While the meal was finishing, he set the table with plates and bowls, napkins and silverware, sliced some of the French bread that Samantha had brought home and rigorously scrubbed the stove. As a final step, he got out a candlestick and placed it on the table and put a red candle in it.

As a finale to the meal, he washed and placed a couple of oranges and apples in a glass bowl and put these and small rounds of red wax wrapped Gouda cheese on the table.

"Something is smelling good in there," Samantha observed. "What did you cook?"

"It's the brisket," Aron responded. "I don't know whether to call it a soup or a stew, but whatever it is it can sit on the stove until you can eat."

"I think that this is winding up," Samantha said. "I am going to need to write up notes from the meeting and send those out tomorrow. I think we covered just about everything we need to do this time around."

Wanting to see how his creation had done, Aron poured out a bowl and sat at the table to try it. He also took some of the rendered marinate and put that in a small glass dish and dipped a point of the French bread into it. He savored the onions and the spices from the marinate that had soaked into the bread. Testing the meat, he found that he could cut it with a fork.

"This is good Sam, come on while it's hot."

"In a minute," Samantha replied. "I am about finished. I'll need to look it over before I send it out, but I think that I have about everything that's needed. We are trying to sell our insurance plan to a company that is now owned by a European conglomerate, and some things that their state insurance offers in Germany we just don't do here. They have a hard time getting their heads around reconciling our private health system with their public one."

"This is nice Aron. Could you get out one of those salad mixes and some dressing?" Samantha requested as she sat down. "I feel like I need some greens."

With Aron munching on an apple and Samantha working on her salad, soup and bread, for the first time that day they had a moment for conversation that was not driven by some pressing activity.

"How was your day?" Samantha queried.

"They came and put in my computers and phone. They did a good job of it and cleaned up after. I haven't had a chance to use any of it yet, but will try it out tomorrow. Sam, don't be shocked when you see what they brought. They said that they had to put in whatever they could buy. Do you remember a band called The Village People? Well, we now have them in our bedroom."

"Say what?" Samantha responded between bites.

"You'll see," Aron asserted. "Enjoy your meal Sam. I think this is an experience that we need to share. It's nearly overwhelming – like being at a concert with them."

Her curiosity provoked, Sam finished her meal, and leaving the dishes on the table the pair walked to the bedroom and opened the door of the darkened room. Aron turned on the light switch and the computer cabinet lit up and the sound of Y.M.C.A. erupted from the cabinet's speakers as the computers powered up. When each new unit activated it added to the neon light show. After a minute the music stopped, the display lights quit flashing and the comparatively subdued sound of humming computers filled the room.

"That happens every time you turn on the light switch," Samantha said. "That's, that's unbelievable."

"I can turn the volume down, but that's it." Aron replied. "The embedded sensors in the cabinet control the computer."

The pair sat on the bed and Samantha asked, "You remember Elizabeth, my sister?"

"How could I forget?" Aron replied. "You remember people who picked ticks off you. I don't know if there is a name for that type of social bonding, but there should be."

"Her high school graduation is coming up, and I feel like I need to attend. You are invited too, if you like. Elizabeth asked about you."

"With our wedding coming up, I need to save my vacation time," Aron replied with a sense of relief that he had a reasonable excuse. "Dad called and he said that he had rented a place where we could have the wedding and reception. Since we can only have 50 people, that hall should work out fine. This will be far less expensive that having the ceremony in the temple and then going somewhere else for the reception."

"We have the date then?" Samantha inquired. "Yes, it will be May 4, and following that we can have the fertility things done," Aron answered.

"And you could not tell me before now?" Samantha interjected.

"And exactly when could I have done that?" Aron replied. "You have been on your calls ever since you walked in the door. This is the first uninterrupted moment that we have had all day, even if I did have to share the news with the band on our wall. Outside of them, you were the first to know."

"All right," Samantha conceded. "I guess that's right. This bringing both our offices to our apartment is going to take some getting accustomed to."

"Sam, I am worried about you getting on a plane, flying down there and being in a graduation and being exposed to this COVID virus.

It's dangerous stuff and could impact the baby." Aron said with a look of concern on his face.

"I've got good masks and will wear them all the time. That is the best we can do." she assured Aron. "If I, or we, both get the virus we are young, healthy and should be like catching the flu. It may even be better for me to have it before I get pregnant to give the baby some immunity. No one really knows yet. I care what happens to the baby too."

A few days later Aron was driving Samantha to the airport to catch her flight to Atlanta. He pulled up to the Delta departure gate. Like she had done many times before, Sam gave Aron a peck on the cheek and with a departing "I love you" grabbed her bag out of the back seat, put on her mask and made her way to security. As a frequent flyer she quickly made her way through security and as a first-class passenger was among the first to board the aircraft.

As soon as she sat down and buckled up she checked her cell phone and found 20 messages of which several were sent from German addresses about the conference they had a few days ago. She decided that she would answer them in flight and put her phone away before the pilot asked.

"You going to the COVID-19 meeting at the CDC?" Samantha's next seat passenger asked. He was a 35-year-old black man with medium-length dreadlocks dressed in a well-fitted light blue pinstriped suit.

"No." Samantha responded, "I have a sister who is graduating from Wilmington high school tomorrow."

"I asked because there are a lot of people from the health industry in the Midwest who are on this flight," he responded. "Delta gave us a discounted rate."

"I work for United Insurance who arranges group Health Insurance policies for larger companies, and this virus has had an impact on our pay-outs and coverage. We are somewhat lucky that most people we insure are younger and have less severe disease," Samantha informed her new seatmate.

"Have you heard how this virus impacts newborns and infants?" Samantha questioned.

"At this stage there are some studies that are started, but there is not a large enough population group to make a good guess," he replied in a serious tone because he suspected that this question denoted more than casual interest. "Are you carrying a child?"

"Not yet, but my boyfriend and me plan to be married soon and we have already started an in-vitro fertilization process because we both have defective genes, something called Haemochromatosis," Samantha replied, somewhat surprised by her own frankness in disclosing this to a stranger.

"I'm in the social sciences," he responded. "One of the reasons for this conference is to set up clinical trials for vaccines and study groups to study the epidemiology of the disease. One of the populations we will be studying are the impacts of the virus on newborns and their mothers. If you will give me a card, I will e-mail you any reports that I hear about and maybe put you in touch with physicians who are studying mothers and newborns. What was the name of that condition again? I want to write it down so I can ask about it."

"It is Haemochromatosis," Samantha responded. "It causes liver damage. It impacts something like one out of every few hundred Americans my doctor said. The disease is restricted to certain white populations from Western Europe."

"Thank you for being so forthcoming," her new-found college said. "If there is anything going on at the conference that can help, I will send it along."

"Thank you for offering to help." Samantha said as she reached into the pocket of her jacket and pulled out a card that he handed across the dividing console into his waiting hand. "I am sorry, but I have some e-mails to attend to."

"This technology business is great, but it has the bad tendency to never let us get away from work," he observed as he pulled out one of his own cards. It read, 'Jose Nance, PhD, Sociology and Human Studies, McGill University, Toronto.' As he handed it to her, he concluded, "However productive this technology might let us be, it is just not healthy to be so closely connected all the time."

"Elizabeth, you look sharp." Samantha said as she saw her sister dressed in cap and gown coming down the stairs at her dad's home.

"I am so scared that I am going to trip over this thing and make a fool out of myself," Elizabeth replied.

"Just take it slow and easy and you will do fine," Margaret told her younger sibling. "I know that you are nervous and all that, but just think how the boys feel. The important thing is not to get all hot and sweated up. There are what, 200 of you graduating today?"

"Something like that," Elizabeth responded.

"Girls, come on, I have the car cooled down, and we need to get you there, "Tom ordered. "Your mother and I are waiting."

"Sam. How was your flight down?" Tom asked as he walked over and gave her a quick hug before opening his arms in a herding motion as he directed his flock out of the door and towards the car.

"It was O.K. Dad, nothing special," Samantha replied as she got into the rear seat.

"Elizabeth, you get into the front seat, so you don't get wrinkled up," Tom instructed.

Always one with a bent towards the practical, Susan, Tom's wife, added from the back seat. "Tom you could not wrinkle this synthetic stuff if you tried. It might be hot as blazes and soak water like a sponge, but it is not going to wrinkle."

"Because of COVID, the ceremony is going to be on the football field and the prom is going to be there too," Elizabeth informed Sam. "The juniors in shop have been making a dance floor and platform for the D.J. They have to get that all torn down and moved before the soccer game on Friday."

Arriving behind a line of cars approaching the football field, Tom unfolded his general plan. "When I get closest to the gate you all get out and find your seats. Elizabeth, you are going to be up front with the graduates and sat in alphabetical order. Susan, save me a seat while I park in the parking garage. We will meet near the exit and leave after the ceremony.

"Elizabeth, this is a wonderful day for you. You relax and you will be fine. Goodbye." With those words he bent over and kissed her on the cheek. He hesitated as a tear rolled down his face while he watched his family walk away. Elizabeth, his youngest, was about to enter another segment of her life. A honk from behind brought him back to reality, and he moved the car into the parking deck. "At least something will be in the shade today," he thought.

Tom saw his family standing apart from the rows of chairs on the football field because the ushers were filling one row before starting another. As close as the chairs were spaced there was no room for a person to walk out once they were in the interior of the tightly packed array. "I hope some old guy doesn't have to pee," Tom thought.

"Susan turned to Samantha and spoke, "I am sorry that Aron could not come. We all would have liked to have seen him."

"From Aron's point of view, you have all seen altogether too much of him," Samantha chuckled.

"I suppose that's right. I can see why he might have thought himself a little overexposed with the tick picking by everyone but the dogs," Tom laughed. "I shouldn't make fun of it, but that was about the funniest thing that I can imagine that a man could be put through on a visit to his in-laws. I'll have to make it up to him somehow - maybe with a vacation somewhere."

"We've got that planned. We are going to Italy for our honeymoon. My Uncle Tyler has been several times and showed me photos of his trips to Rome, Venice, Pompeii and Sicily. I've got plenty of airline and hotel credits so it won't cost us hardly anything."

Taps on the microphone signaled the start of the ceremony. There were speeches that were thankfully brief from the Superintendent of Schools, the Class Valedictorian and the Principal as row after row the graduates rose and proceeded across the stage to receive their diplomas. A thunderstorm was building up and a cooling breeze swept the field. In fear of a general wetting, the announcements of the names were speeded up to the extent that the graduates were trotting across the stage as photographers and parents tried their best to capture this fleeting moment with their cameras and smartphones. Ultimately Elizabeth's name was called, and she turned briefly to the front to allow a photo and then scampered off the stage to her seat. By the time they came to Zorag, everyone was relieved that the ceremony was over. Class members hugged and chatted while the parents attempted to make their get-a-ways as efficiently as possible.

Feeling like a bull elk gathering his cows, Tom spotted a figure he thought was Elizabeth amid the throng of identically clad figures and tapped her on the shoulder, but it was not her. He gave his apologies and looked further. Walking back toward the gate he saw that Elizabeth had already rejoined her family.

"I am glad and sorry at the same time that this is over," Elizabeth remarked. "I had a lot of good times at this school and it's hard to say goodbye to everyone knowing that I will never see some of them again."

"You will see most of them tonight if it does not rain you out, and there will be class reunions from time to time," Tom said.

Once in the car Tom said to Elizabeth. "Every boy in that class wants to get laid tonight, and likely doesn't care with whom. If that is going to happen, you bring him to the house and use your bedroom. Here are some condoms to put in your bedside table. He can stay the night and join us for breakfast. If he is man enough to do that, then he may be worth having."

"I know Dad." Elizabeth replied. "You have told us many times."

Returning home, Samantha and Margaret helped Elizabeth with her hair and prom dress while Susan made some sandwiches. Because of the outdoor location there was to be no food served. No alcohol was allowed, but everyone knew it would be there anyway. The freezer chests of canned drinks and a bubbling punchbowl of lime and dry-ice punch were the only things offered to the prom goers, faculty chaperones, off-duty police and band.

Putting the keys purposefully in Elizabeth's hand," Tom said. "I am going to lend you the car. Do not let anyone else drive it. These boys think they are invincible and each year somebody gets killed in some senseless traffic accident. Have a good time. Bring the car back in one piece, and bring yourself back safely."

The after-prom vigil begin with everyone remembering the worst events of their own proms as they sat on the chairs and sofa in the living room and watched "Flashdance" on TV. That was followed by a documentary about some castle in the U.K., and on and on as the hours slowly passed.

"Was it like this on my prom night?" Samantha asked.

Grasping his Old Fashion in his hand, Tom replied. "Much the same. We wait, we pray a little, think about all the things that could go wrong and sit, sit and sit. I never did the shotgun thing waiting for some boy to show up, but that was among the options that I considered."

At 2:00 AM a pair of headlights struck the front of the house, and Margaret went to the window and peered through the shades.

"Is she alone?" Tom asked.

"I can't tell. Yes, she is alone," Margaret said.

"Then everyone to bed," Tom ordered. "No questions tonight. She will tell us what she wants us to know in the morning."

Although the lights were still on in the living room when Elizabeth entered the living room only a half-drunk glass of her dad's Old Fashion with ice still in it remained on a coaster on the side table. She took a sip, made a face, and walked over to the kitchen to see what was left in the fridge. She grabbed a sandwich and dumped the rest of the drink down the sink.

Chapter 7

Beth and Larry

Startled by a rap on the door at 6:00 A.M., Samantha woke to Tom's announcement, "Beth and Larry are coming for breakfast. They have something important to tell us." Samantha listened as the same series of raps and words was repeated at Elizabeth's door across the hall.

"Well, I have an announcement too," she thought. "If they are going to tell us what I think they are going to tell us, we might as well make this a double hitter." She had not said anything because yesterday was Elizabeth's day, and she did not want to detract from it. Her flight was not until 3:00 PM, so she just put a robe over her bedclothes and went downstairs.

She met Elizabeth coming out of her room and asked, "How did things go last night?"

"Generally miserable. I'll tell you all about it at breakfast," she replied.

Susan was at the stove frying up some Canadian bacon and eggs while tending a pot of bubbling grits. "Elizabeth, Samantha set the table right quick and pour everyone some coffee, this will be ready in a few minutes."

Putting the steaming platter of food on the table and taking the toast out of the toaster, she said, "Everyone sit down and eat while it is hot. I'll cook some more bacon and eggs when they get here."

"Elizabeth, how did your night go?" Tom asked.

"Just miserable, Dad, miserable," Elizabeth began. "Some rain got on that floor they put down and it was too slick to dance on, the heat was terrible, everybody was dripping with sweat and the bugs really came out after that shower. Everyone just stood around while the D.J. tried to get his sound system working, and then people just left. I took some of my friends to the Sonic and the I-Hop and we sat around and talked until they ran us out. I took everybody home and then came back here."

"Were there any incidents at the prom," Tom inquired.

"One of the boys took his coat and shirt off and ran around trying to kiss the girls and get them to kiss him on his nipples," Elizabeth related with disgust. "The school police took him home. If anything else happened, it occurred after we left."

"I am sorry that things turned out so badly for you," her mother replied. "I think that we could have had a better function here."

The doorbell rang and Tom got up to answer it. "Beth, Larry come in, come in and sit down. Mom will make you up some fresh eggs and bacon."

"I am sorry to make everyone get up this early," Beth apologized. "This was the only time that we would all be together. Larry and I are going to get married next week.

"Elizabeth, I am sorry that I missed your graduation. I would have liked to have been there. It is just with my special needs kids, I can't get away on a weekday."

"What's the rush to get married?" Tom asked.

"It's the job," Larry began. "I am doing the electric design and installation of the electric circuit at an automated warehouse distribution center in South Dakota. This will be the largest such center ever done. With this COVID stuff, I may not be able to get back for months. We decided to get married now, so at least we could set the date and do it."

"You can't even get home on the weekends?" Tom asked.

"The time costs on this job are enormous," Larry said. "Every day's delay runs in the hundreds of thousands of dollars. There is a tremendous amount of concrete to pour and over 10 miles of wiring in the building. I have only one shot in making sure that the right conduits and cables get put in the right place and we are scrounging from all over the world to get them. Usually this is fairly simple. You contact your suppliers early enough, and they deliver what you need. But with factories shut down in Asia, particularly China, and shipping delays, it is really tough."

"Well I have an announcement too," Samantha began. "Our wedding will be in Detroit on May 4. Aron's dad has a reception hall booked. After our honeymoon in Italy, we are going to have the implantation done and expect the baby next February. We can only have 50 people, but all of you are invited. We'll get tickets, fly you up and arrange for places for you to stay."

"O Beth, Samantha, I am so very happy for you both," Susan said. "I think that I'm going to cry." Susan left her stove and went to hug Beth and Samantha while Tom rose and took over scrambling the eggs.

"These are ready." Tom announced.

"Sorry," Beth said as she looked at her watch. "I have to leave. I need to get to school as close to eight o'clock as possible. I have a meeting to explain how I am to attempt remote learning with my six students. Some of the participants will be there, but most will be on Zoom.

"After that, I have to select and test computers and phones that the school is going to deliver to the parents. I have written up pages and pages of stuff for each child, but I have no idea if any of it can work. I don't know if the parents can install the computer links or where they would put them. Oh, and everybody there has to put in their two-cents worth whether they know anything about the kids or not.

"Larry can stay if someone can drop him by the apartment."

"We'll get him back," Tom replied. "You go ahead and make your meeting."

"Thank you, Dad," Beth replied. With a "love you all" she rose from the table and ran out the door attempting to save as many seconds as possible.

A silence ensued after Beth's departure as the momentous

announcements that were just made sank in for those sitting around the table.

To break the tension Larry turned to Samantha and inquired, "What's Aron like and what does he do. I have never met him, and it looks like we are soon to be brothers-in-law."

"He is a city boy." Samantha began. "A Detroit Jew who has never really been out of the city. He's tall, about 6-feet 7-inches, fair skinned and has brown hair. We met while I was giving a presentation at University Hospital where he works and things progressed from there."

"He's a doctor then?" Larry inquired.

"No. He assists those using the fertility clinic at the hospital in filing for their insurance and arranges grants for those who cannot afford critical care for mothers and newborns," Samantha replied.

"One thing I don't understand. Why are you going to have an implantation?" Larry earnestly inquired.

"We both have an inherited gene defect," Samantha began. "It is called Haemochromatosis, and if both parents have the same defect the child can have liver and other potentially fatal diseases. This condition impacts about one in every several hundred Americans. It is more common in Western Europe."

"If this is inherited, can Beth carry it too?" Larry asked with a note of concern in his voice.

"It's possible," Samantha responded. "This defect is not passed down to all of the children and having only one copy of a damaged chromosome does not often cause problems. It's when a child gets copies of the worst mutation from both parents that difficulties result, as I understand it."

"We'll both get tested," Larry responded with a hint of earnestness in his voice. "I am sorry that you and Aron have to face this."

"You'll need to make a special request for this blood exam in the U.S. It is not done with the usual tests for STDs before a couple is married," Samantha informed her future brother-in-law.

"Larry, tell me more about this automated distribution center?" Tom requested. "This sort of thing sounds like it could replace thousands of workers."

"That is exactly the case," Larry responded. "The general plan is to have automated machines called 'ants' run on tracks to pick up items from bins and move them to other stations where they are packed, put in individually made boxes, labeled with the shipping address and taken to the correct door for the next pick up. Each of these operations was done by workers, and now the plan is to replace them with robots. This system can run 24-7 and be managed by 20 to 50 people instead of hundreds of shift workers. This one operation is supposed to replace four manual distribution centers. Putting in this center will cost millions, but this money will be saved from wages and benefits that would be paid to workers."

"And this is going to be the first one?" Tom asked.

"Not exactly the first, but the first on this scale. Some of the components have been used previously, but others, like this design of 'ants' have not been used before. They are electrically powered. As they move on rails, they self-generate a portion of the energy that makes them run. This facility is designed to start with 50 of them with a work plan that 45 would be available at any given time while the others would be in maintenance."

"How are you and Beth getting along?" Samantha queried. "Did she tell you about her and Aron?"

"Only that she thought that he was a good-looking guy and regretted causing you any difficulties," Larry responded. "She told me that she threw his suitcase out a window."

"Yes, she did," Tom replied. "Some bottles broke, and we had to wash all his clothes and borrow others from a neighbor for him to wear to visit Samantha's stepfather's parents. His wearing those clothes almost got him into trouble at a truck stop, but that turned out all right."

"How did that visit to your stepdad's parents go," Larry questioned Samantha, considering that this was a trip that he and Beth also needed to take.

"Don't make love in the pasture," Margaret replied. "They came back covered in ticks and everybody had to help getting them off. Up here I worked on Aron and Elizabeth picked them off Samantha. To say that we got a good look at him would be an understatement."

"I imagine he was embarrassed," Larry postulated.

"He was shy, even around me," Samantha said, "But I think that he is over all of that now."

"Good," Tom said. "I was thinking of something to give him for a wedding present, and I think that I have found the ideal thing. That is a week at the All-Gospel Bible Camp and Nudist Colony in Terlingua, Texas. They put people out where an enormous volcano blew up and left a caldera called The Solitario which is in a state park. They give them sneakers, a blanket and a spear and leave them alone for a week. They have to find water, food and make shelter and fire. That is just the thing a city boy like Aron needs to broaden his experiences. It's a real test of manhood."

"Dad," Samantha interjected, "Where did you get such a crazy idea. I want to marry this man, not kill him. He would not have a clue how to survive outside a city.

He has many desirable qualities, but there is not an iota of 'Daniel Boonism' in his character."

"If you are going to have his child, I guess that we need to keep him around," Tom agreed.

"Larry, I'll take you wherever you need to go. Margaret has to leave for her first class fairly soon. Then I have to come back and take Samantha to the airport. One never knows about traffic, and I want to allow plenty of time. Sam, if I have any problems, I'll phone in time for you to catch a cab."

During the four years that Larry and Beth had been living together Larry and Tom had casually talked at family functions, but now that the couple was going to take on a life-long relationship, Tom felt that some more serious questions were in order.

Once in the car Tom began, "Larry I know that you are working for a specialty firm that hires you out as a contractor. What do you see for your future?"

"This particular job is an excellent opportunity for me," Larry explained. "It pays really well, and I do not have the expenses of maintaining an office or having employees. I am putting away a lot of my salary and ultimately will start my own virtual firm."

"How would that work?" Tom inquired.

"All of the typical business services that I need like an answering service, booking, shipping, billing and collections can be contracted out. I share these services with perhaps a dozen other businesses and can continue to work out of my home office, much like I do now. If I need a room for a conference, I can book one downtown."

"I can see the income side," Tom probed, "but what about health insurance?"

"Beth's job as a teacher will cover us both after we're married. I have been carrying a catastrophic policy, which is relatively inexpensive. This COVID-19 business has really scared us both. Beth got it at school, and I got it from her. We both had mild cases. Now that schools are closed, she will be dealing with her kids remotely, but she is fearful that the gains she has made over the past few years are going to slip away.

"One day I hope to make enough that she can quit that job and we can start think about our own kids. Right now, it is like she already has six that she is parenting, and to take on another child once she comes home is just more than she, or frankly me, can do right now."

"I know that she really gets mad when things don't work out with those children," Tom agreed.

"It was like that the night that she confronted Aron," Larry said. "He caught hell from her for something he had absolutely nothing to do with. I don't know him, but I certainly feel for him."

"Could that offer for a trip to the Terlingua Bible Camp's Solitario experience be extended to me," Larry inquired. "After I get through with this job, being somewhere alone for a week sounds good. My father lives in Texas, and I know West Texas from living on the ranch. Perhaps Aron and I could do that together after his baby comes and this contract job is over."

"I see that as a reasonable suggestion," Tom stated. "We have time, nine months at least, to work on it. I believe that it would be good for you both."

"Has he had any outdoor experiences?" Larry questioned.

"Only a couple that I heard of. Many families keep lake cabins in Michigan. I don't think that his ever did, but he and Samantha went up with someone else. When they got there the cabin had been lived in by a family of raccoons. Everything was torn up and smelled terrible. They spent almost their entire stay helping the others clean up everything and after washing everything in the lake and using a lot of Pine-Sol they finally got everything scrubbed down. He did catch some pike, and they ate those. Somehow on the way back or from something he ate up there, he got the shits, and felt generally miserable from the experience.

"The other time Samantha convinced him to go camping. They got set up and went inside their tent, and then it started to rain so hard that their tent flooded, and they pulled out wet and miserable in the middle of the night and came home. They also met a porcupine on that trip who tried to come into the tent with them. I guess that he didn't like being out in the rain either."

"I can see why he would not be enthusiastic about spending more time in 'The Great Outdoors.' These must have been unsettling, and he did not even meet any coyotes, wolves, bears or mountain lions like I have.

"It looks like it is going to be a 'hard sell,'" Larry concluded, "but I would like to have him along. By the time we both get through our life issues, I think that almost any change will be welcome."

Chapter 8

A Wedding

HIS HOUSE BUSTLING WITH SIX bridesmaids attempting to get dressed using the house's three baths, Fred eased his way past two of them to see how Beth was progressing towards getting ready for her wedding. Thankfully Larry was at Tom's with his groomsmen, and he had only the female side of the operation to deal with.

Fred asked, "How are things going in there?"

"It's my hair dad, it just won't do right," Beth replied.

"Believe me," Fred replied with a tone of exasperation in his voice. "Larry really won't care if you have any hair or not. The bus will be here in five minutes, and everybody needs to get on board. Larry's parents are here from Texas, and they are already on their way to the church."

"We're hurrying as fast as we can dad, this is not like pulling on a pair of jeans and a top," Samantha replied in Beth's defense. She had arrived the previous night, but Aron remained in Detroit. Despite her attempts to convince him otherwise, he asserted that he wasn't really a family member yet, and did not want to take a chance at traveling considering the increased severity of the COVID outbreak. He'd come down after they had their baby, but for now he would stay home.

"I'll blow a whistle when the bus gets here but everyone be careful about going down the stairs. No one needs to fall and go to the wedding with a busted nose," Larry informed the throng.

"And husband dear, how do you know that would happen?" Lollie asked.

"Call it a premonition." Fred replied. "With all this turmoil anything could happen. Five-minute warning everyone. When you are ready come down to the living room and have a seat."

From a guy's viewpoint it would be comforting to know that things were going more smoothly at Tom's on the other side of town, but that was not so. Margaret and Elizabeth were at Fred's house with the bridesmaids. Rover was not happy with this invasion of six strange men, and Tom had to take him to a nearby vet to be boarded out in luxury pet accommodations. Returning, he had to quickly get dressed to reunite with his family and get everyone to the church. Once dressed, he lined up the groomsmen shoulder to shoulder, adjusted their ties and helped Susan pin the boutonnieres on their coats.

"Fellows, you were all at the rehearsal and know what to do. When we arrive, you are to act as ushers and guide the families to their sides of the church and sit them down. You are going to be in close contact with people so wear your masks. I will tell you who to seat, where. I want to keep two empty isles between the older members of the family and the rest of the audience. We want the members of the audience to be dispersed in the church. Offer everyone masks and help them put them on if you need to.

"Once the processional starts you escort the bridesmaids to the altar. Beth will follow alone and when she stands beside Larry you and the bridesmaids do an 'about face' and return to your pews at the back rows. You will be sitting behind the bridesmaids. When the ceremony is over and the pictures are being made, you escort the attendees out of the church. When only the family members remain, you escort them out so they can be on the steps if they wish to be there when the bride and groom leave the church, and more pictures will be taken. If some want to go directly to their cars help them. It is going to be hot, so here is a handkerchief for each of you.

"If you need to, or think you might need to, use the bathroom, go and do it now. It is going to be a half-hour to the church and another hour at least at the wedding. I don't know many of you, but we'll have a chance to get acquainted at the reception. Thank you all for coming and helping out," Tom concluded.

A blast from Fred's whistle signaled that the bus had arrived. Beth's bedroom door opened, and she came out dressed in a flowing white gown studded with synthetic pearls with a train and vale covering her brown hair. Samantha guided Beth's hands to the stair rail and Beth in her high heels proceeded step by step down the stairs.

"Hold it for a second Beth," Fred requested. "I want a picture." Raising his camera he took his photo and the flash somewhat blinded Beth.

"Beth, wait for a moment, and then come down," Fred commented. "I don't want you to fall." Unexpressed, but running through his mind was the story of a bride on her wedding day who stood too close to an open fireplace and had her dress catch fire. She was burned to death and buried the day she was to be married. He did not want any similar events to mar this generation's happiness.

Now that Beth was in the living room, Fred and Lollie helped each of the bridesmaids through the bus door and into their seats. Samantha went in and was followed by Beth.

"Everyone here?" Fred asked. Receiving no reply to the contrary, he nodded for the driver to proceed, and this detachment of the wedding party was on their way — something he had serious doubts about a half-hour before.

The groomsmen's buss was the first to arrive and as they got off the bus an ashen-faced Albert Hopkins told Tom, "I feel terrible, I think I am going to throw up."

"The bathrooms are in the basement. Go." Tom urgently requested. "You," grabbing the next groomsman getting off the bus by the sleeve, "help him."

Assisted by his new-found buddy Harvey Winesap, Albert stumbled up the long flight of steps to the sanctuary door, caught his foot on the lip of a step, fell on his face and started a profuse nosebleed. Harvey pulled Albert up and shouted, "Where's the bathroom," and helped to support him as they entered the vestibule and headed for a door on the left-hand side. Holding the spring-loaded door open while Albert entered, Harvey caught him by the collar to keep him from falling down the steep spiral stairway, and in the process almost strangled him.

At the bottom of the stairs Albert saw the bathroom door and breaking free made a dash for it holding one hand over his mouth while opening the door with the other. He headed for the first stall and yanked open the door while violently puking over a startled older gentleman sitting on the stool.

Realizing the enormous nature of his social infraction he muttered, "Sorry, couldn't help it," and proceeded to franticly unbutton his belt and loosen his trousers while moving to the next, thankfully empty, stall.

Harvey stood outside and listened while the explosive sound of gas and fluid release erupted accompanied by smells that were so rank that he started to feel that he might have to throw up himself.

"Are you all right in there," he inquired.

"No I am definitely not all right, and my clothes are ruined," came the reply from the other stall. The occupant was Larry's father, Larry Austerhouse Sr.

"I'll get Beth's father and see what we can do," leaving them both in their respective states of misery, Harvey ran up to find Tom and inform him of this predicament.

"Larry," Tom said, "Your father needs a shower and a change of clothes. You find the pastor and see if there is anywhere in the church where he can shower and any clothes that would fit him. We don't have time to send him back to the hotel."

Proceeding down to the bathroom, Tom, stood in the middle of the floor and addressed both participants of the most recent disaster.

"Mr. Austerhouse, I am dreadfully sorry about what happened. I am afraid that you might have been exposed to COVID, and we are looking for you a place to shower and some clothes for you. Larry is talking to the pastor.

"Albert, is it?" he enquired.

"Yesss," came a weak reply from the other stall.

"You stay down here until things are over. We'll get you to the hospital if you feel like you need to go."

"I feel a little better now, but I don't know if this crapping and puking stuff is over or not," Larry responded. "I have never felt so bad in all my life."

"All right then, I'll call 911 and get an ambulance," Tom responded. "At least they will be able to treat you in the emergency room."

A few seconds later Larry Jr., breathless from running back and forth through the length of the church, came in.

"Dad, I talked to the pastor. There is a shower, and he said that he would freshen up some clothes that you can wear," Larry Jr. informed his dad. "He is a big guy and nothing he has will really fit, but we did find some outfits that are about your size."

Fred and the bridesmaid's bus arrived at the church. Fred and Lollie helped the bridesmaids off the bus and up the long flight of stairs into the vestibule of the church. The groomsmen, now directed by Susan in Tom's absence, came to help get the ladies into the relatively cool confines of the back pews of the sanctuary.

Noting the presence of only four groomsmen rather than six, Fred asked Susan, "Where is Tom and the others?"

"One of them got sick," Susan replied, "Tom and another of them are down in the bathroom. They've called an ambulance. Larry's father got puked on and they are getting him showered and finding a change of clothes for him. So far as possible, everything is being taken care of."

Walking up the steps of the church Tom stepped around a pool of blood and string of droplets tracing Albert's dash to the bathroom. "I said there would be blood, I knew it. I just didn't know whose blood it would be."

The relative calm of the moment was interrupted by the arrival of the ambulance. The EMTs put cones and yellow tape around the pool of blood on the steps and retrieved an obviously weakened Albert from the bathroom. They supported him as he walked down the church steps where he was placed on a gurney.

Relieved to be removed from a terrible situation, Albert lifted a hand and gamely waved to those standing on the porch. Whatever happened later, he would be thankfully removed from the situation.

The ambulance pulled away and a since of normality returned to the church.

The organist started a Bach prelude, the guest sat silent in the seats and the procession to the altar was being organized by Fred and Tom.

"Larry where is your father?" Tom asked.

"He will be here in just a second." Larry Jr. replied. He got showered and they found clothes for him. That Neman Marcus suit that he was wearing was trashed. They are going to burn it."

"Larry, this is supposed to be yours and Beth's day," Fred added. "Are you all right?"

"Nervous, a little, I guess," he responded. "But with all this shit coming down, I really haven't had time to think about it."

"Go on down and take your place at the altar," We'll send your dad down when he gets here.

On the other side of the isle Lollie and Susan were equally solicitous about Beth.

"I don't know why, but I am starting to hyperventilate," Beth replied.

"I have just the thing," Lollie said as she reached into her bag and pulled out a paper sack. Put this over your mouth and breathe in and out of it slowly. Everything is going to be fine, just like the rehearsal."

The organist increased the volume of his music as robe-clad figure carrying a staff joined the men. It was Larry's dad wearing the only clothes that they could find that would fit him. He was dressed as a biblical shepherd in clothes typically worn during the church's Christmas Passion Play.

"It was either a shepherd or Pontus Pilot," Larry Sr. whispered to his son. "I thought the shepherd would be more appropriate."

Sir, we are short a groomsman," the Wedding Director asked. "Would you escort Samantha to the altar and then you could take your seat beside your wife in the front pew?"

Larry Sr. instantly grasped the twin symbolisms of his being a shepherd leading the flock of groomsmen and of his guiding Larry Jr. through life.

"Sure, why not?" he assented. "It might be a bit unorthodox, but what about this wedding hasn't been."

With everyone in place, the wedding director signaled to the organist to begin the processional and with "Here Comes The Bride" the twin lines of groomsmen and bridesmaids began a measured walk down the aisle. While everyone's eyes were focused on the processional, the parents went around the outside of the seating area to their places in the front pews and Tom took his place at the altar. When the volume of the music increased, Beth begin her solitary trip towards wedded matrimony.

The bridesmaids were wearing identical pink dresses that incorporated much less voluminous skirts than the bride's outfit. The groomsmen were dressed in identical blue suits, cummerbunds, vests and jackets. When the group reached the altar, Larry Sr. took a seat in the front pew next to his wife. Helga gave her husband a startled look.

"I'll explain later," Larry Sr. whispered.

When Beth took her place beside her future husband, the groomsmen and bridesmaids returned to the back of the church where they took their seats.

Rev. Dr. White had shortened his usual ceremony. After relating the story of Adam and Eve and admonition to "be fruitful and multiply," he quickly got to the vows.

"Do you, Larry Austerhouse Jr., take this woman, Beth Williams, in the presence of these witnesses and this company to be your lawful wedded wife, richer or poorer, in sickness and in health, until death do you part so help you God."

Larry swallowed hard, turned to look and Beth, and replied, "I do."

Beth's "I do." was delivered a bit more forcefully than Larry's response, as if after a long battle a fish was being reeled into the net.

"You may give the rings," White intoned.

A patter of feet was heard from behind. Larry Jr.'s sister's youngest daughter, Katy, likewise dressed in pink, moved from her seat in the front pew with a red velvet pillow with two rings sitting on top.

Rev. White explained to the couple and the audience that these two rings represented the circle of life and the renewal of life and faith from generation to generation.

Finally, he concluded, "By the power vested in me by the holy mother church and the State of Georgia, I now pronounce you man and wife. You may kiss the bride."

Larry lifted the vale and kissed Beth with a feeling of relief that this event was coming to an end.

Katy, once ring bearer, now flower girl, took the ribbon-wrapped bouquet from her mother's hand and prancing up to the alter presented it to Beth who accepted it with her left hand while she held Larry's hand with her right.

Flashes erupted from cameras and smartphones as people attempted to record the event. As they were being escorted from the church, the wedding photographer set up his camera and posed the couple for a variety of shots. When he was finished, he directed that they move to the church doors for another round of photography.

As things were winding down a darker skinned elderly black-clad woman also climbed the stairs intending to go into the church to pray.

While the couple was receiving congratulations under the church's portico, she went up to Larry motioned for him to bend over and dipping a finger in the pool of blood on the porch gently put a dot of it on his left cheek.

"Blessed," She said as she crossed herself. She then motioned to Beth and did the same.

"What was that all about?" Larry Jr. inquired.

"She is from somewhere in South America," Rev. White replied. "This is the closest church to her house and she comes to pray. Obviously Catholic, she means no harm, and we accept her as part of our congregation."

With all of the formalities of the ceremony done, a couple of traditional tasks remained. Beth, poised with her bouquet prepared to throw it at the clutch of bridesmaids below. A wind gust lifted the bouquet out of reach of the bridesmaids. More out of self-defense and instinct rather than intention, it was caught by Harvey Winesap, whose face instantly turned red. He attempted to give it to Samantha, then Margaret and then Elizabeth who refused with a general round of laughter from all. Stuck with something he could not throw away but did not want to keep, he resigned himself to keeping it with the idea of someday returning it to Larry Jr. and Beth.

Rather than rice peppering the couple as they walked down the steps, they were showered with dry mini-footballs of Quaker Oats which the sparrows and pigeons would find more digestible.

Chapter 9

Consequences

"Whatever we do," Samantha told Aron when she arrived back at the apartment in Detroit, "I don't want a wedding like that. That was too much trouble, too many people and too much bother."

"As I understand it from Dad," Aron replied. "There will be no more than 50 people and the wedding and reception are going to be in the same place with one following the other. Maybe we could have a few people over at Dad's or here, but that is going to be about it. If you want to change something we can, but that is the way it is for now."

"From what I am hearing, things are going to absolute crap down there," Samantha reported. "That guy that they took to the hospital was diagnosed with COVID-19. Larry's dad started showing symptoms as soon as they got back to Dallas, and he was taken to a medical center where he and his wife both tested positive. They are both being treated. They don't know when they will be able to go back to their ranch."

"Sam, how are you feeling?" Aron inquired with a note of concern.

"Not so hot," she replied. "I am starting to have a cough and feel like my throat is covered in mucus. I also have something of the runs. Like right now!" she interjected as she made a dash to the bathroom.

Going into the bedroom, Aron fired up his computer to check on

COVID-19 symptoms. The driving beat of Y.M.C.A. and the light show did little to comfort his anxiety. What he found was a pages-long list of symptoms and no way to cure the disease. He found recommendations which encouraged home treatment with over-the-counter drugs unless conditions became life-threatening, or the patient had preexisting conditions.

There was nothing that he saw as being helpful, except to lay up a stock of cough syrup, some anti-diarrheal medicines and something for joint pain.

When Sam returned Aron asked, "Did you ever hear anything from that black man that you met when you flew to Elizabeth's graduation?"

"No. Not yet," Samantha replied.

"Give me his e-mail address." Aron requested. "I will contact him and maybe he can give me some better information. I don't want to ask anyone at the hospital. I definitely will not ask in a hospital e-mail that could be tracked to me. They might do anything, even lay me off. I just don't know."

"I've got his card," Samantha said, and she dug into her purse. "Here it is. He is Dr. Jose Nance at McGill University."

"I'll write him an e-mail telling him that you might have COVID, and find out what he can tell us," Aron offered as he switched on the light to be greeted by music and lights of his communications devise. A few minutes after sending the e-mail, he received a reply and a Canadian telephone number to call back.

Aron dialed the number and Nance immediately picked up.

"Sorry to have to have you call back, but any information about COVID-19 is considered so sensitive that I dare not put it in an e-mail."

"I understand," Aron replied. "I am Samantha's boyfriend. She went to a wedding and she and others came back with COVID. She had a dry cough, is sick to her stomach and tired; and wants to know what to do?"

"I wish that I could tell you something definitive, but I cannot," Nance informed. "Some small study populations among pregnant women and those with newborns have been started. Data is coming in, but it is such a small data set, what the doctors are telling me is little more than wishful thinking. Some have said, with very sparse data, that having the disease before delivery may give the baby some natural immunity."

"If that is true, why don't they say so?" Aron asked.

"There is so much that is not known about this virus," Nance continued. "It appears to be rapidly mutating. Is this virus the only variety? Why are there such a wide variety of symptoms? What is it about the virus that causes some people to have much worse disease than others?

"No physician is going to suggest that anyone be exposed to a deadly disease in order to prevent transmission of the disease to a newborn until there are rock-solid studies from large populations to confirm it, and it passes muster in a peer-reviewed medical journal. Everybody knows that this worked with Smallpox, but no one is going to put their heads on the chopping block about this one until they know a damnsite better about what they are dealing with."

"Can you give us the name and contact information of the person who is directing this aspect of the study?" Aron asked.

"Sorry, I can't do that either," Nance replied. "There is so much fear about violent political backlash, that the only people authorized to release such information are the national heads of organizations such as the CDC, NIH, Provincial and State health departments. The researchers need to be able to do their work, not be bombarded with thousands of telephone calls from people desperate for information. When I hear that someone is going to release something, I will send you an e-mail," Nance concluded.

Returning to the kitchen Aron told Samantha, "He was concerned, tried to be helpful; but basically said that no one knew that if having the virus before you were pregnant was helpful or harmful, and that an answer would not likely to be coming any time soon.

"Sam, we are on our own for this one," Aron summed up. "We are in this together. Whatever you do I will do. If we both have it at we can look after each other for a week until this stuff runs its course. After we are sure that we are both over it, we can get married and complete the conception."

"I love you Aron," Samantha replied.

"Me too," Aron responded as he took her in his arms and gave her a deep kiss with his tongue entering her mouth.

"Tom, what have you heard about everybody?" Susan inquired.

"Albert, who was sent to the hospital, is ready to be released; but he doesn't have anywhere to go. He is a med student at Tulane, but they have closed the university. He's an orphan and doesn't have any family. I feel like I need to go get him and bring him here and let him live with us until things can get sorted out."

"I suppose that would be all right for a few days, but with what they are talking about he could be trapped here with us for some time with nothing to do," Sara replied. "What about Larry Jr.'s parents?"

"His father had to be put on ventilation, but came out of it Larry Jr. tells me. He is getting better, but is weak. He has housekeepers and ranch hands at his place in Texas, but he and his wife may need professional nursing until they recover."

"Can Larry Jr. help?" Susan inquired.

"He said that he would like to, but can't get away from that warehouse job he is doing in South Dakota. He said that this was considered 'critical infrastructure' and this work was authorized to continue despite the pandemic. He is going to be stuck there for months. He said he was sick, and Beth was sick, but they were getting over it – much like Margaret and Elizabeth are. I guess sooner or later, everyone is going to come down with this junk, including us."

"As Albert has at some medical training, maybe he could go to Texas and help look after Larry's parents until things sort themselves out," Susan suggested.

"That might work, but it all depends on what Albert, Larry and his parents have to say about it. I'll pick him up this afternoon and we will see what develops," Tom concluded.

Albert was glad to see someone he recognized pick him up outside the hospital. Tom had not been allowed to visit or even come inside the building. Except for periodic telephone contacts with his friend Larry

Jr. and more recently Tom, he had gotten through this hospitalization alone.

Riding back with Tom, Albert went through a bit of his life story. "My parents, who were not well off, were killed in an automobile accident when I was 10. I was put in the state orphanage in Louisiana and because of my age was not adopted out. I won a scholarship to Tulane, but when I turned 19 my support from the orphanage stopped. I have been working jobs trying to pay my expenses, buy books and cover other stuff that I needed. Not having any parents or assets, I could not borrow any money and had to scrape buy by myself. Thanks for putting me up. It has been so long, I don't know if I know how to live with a family anymore."

"Albert, that is all right. We have a spare bedroom upstairs that you can use, and I can set you up with a computer station so you can do something with your studies online. The only problem that you are likely to have is with Rover, our dog. He takes a little time to warm up to men he doesn't know. He is very protective of the girls and might not let you near them until he gets use to you."

"I am going to have my own room?" Albert replied excitedly. "That has never happened before. I was in dorms at the orphanage and in college, and never had a room by myself. And maybe a dog? That is just too much. I do not know how I can ever thank you.

"How are Larry's father and mother doing? I heard that they were taken to the hospital in Dallas."

"They are doing better, but are going to be sent back to their ranch. They are going to need some medical help, and Susan and I were thinking that you might go and stay with them until they recover."

"I feel terrible that I was the one that started all of this," Albert replied. "Although I don't know them, I feel an obligation to help them if I can. Maybe I could get some sort of in-service school credit for it."

"Before you agree, I need to tell you what you are getting into," Tom responded. "Larry Sr. made a pile of money in the Dallas real-estate market. His father and his wife's father owned adjoining ranches in West Texas near a small town called Study Butte. When he sold his Dallas holdings, he returned to the ranch to live full time. These are nearly barren rocky ranches where in good years they can raise a few cattle, sheep and goats. This is nothing like Louisiana.

"The people who live there are just as hard as the country they inhabit, and don't take any shit from anybody," Tom began. "They are good people, generous to a fault to their friends; but implacable enemies who hold grudges for generations. Everything out there wants to eat, bite, prick or wound you. You can't step outside without being in danger. Larry Jr. spent his later teen-aged years on the ranch and loved it. You may have an entirely different experience."

"You have certainly given me something to think about," Albert responded. "Larry Jr. told me something about the ranches, and I think that if he could live there I could too, although it would take some personal adjustments."

"I'll set up a conference call with everyone and see what can be worked out," Tom concluded as he drove up to his house.

"This is a very nice house," Albert observed.

"I had it built specially for us," Tom said, "It has some unusual features."

Walking into the back door, the pair entered the kitchen and were instantly greeted by Rover who initially bristled and growled at Albert.

"Rover," Tom said. "This is Albert. He is going to be staying with us for a few days." Taking Albert's hand in his own he pushed it forward towards the dog's nose for a sniff. Rover sniffed the hand and then stepped back and sat down.

"That is all the introduction that we need for now. We will take these getting-acquainted steps in stages. Come in and I will show you the bathroom."

"This is your bathroom?" Albert asked. "With that hot tub it looks like something from a rehab center. I'm sorry if I made an inappropriate comparison, but that is how it strikes me."

"You are perhaps more nearly correct than you know," Tom explained. "We as a family hot-tub together and have ever since the children were small. I wanted my girls to be introduced to life and sex in a safe way in this house and not in a back alley or some parking lot in town."

"You mean they did it here?" Albert asked with a note of doubt in his question.

"Not exactly," Tom started. "I have four daughters. My two oldest by my first wife, Samantha and Beth, live with Fred and Lollie across town, although occasionally one may spend the night here. Their half-sisters Margaret and Elizabeth live here. As you know Beth married Larry and Samantha is to marry Aron. Margaret is attending college and Elizabeth just graduated from high school. She will start college next year.

"Margaret did it with her first boyfriend upstairs in her room, but Elizabeth hasn't yet. You may or may not be attracted to each other and may or may not sleep together. That is up to each of you to work out for yourselves. Both are on the pill and there are condoms in the bedside tables. I insist that you use them."

This living arrangement was altogether out of Albert's wildest imagination. With a gulp and a lump rising in his throat, the best reply he could muster was, "I see."

Going up to the bedroom, Albert put down his suitcase and said to Tom, "These are all that I have with me. They're dirty, and everything else I own is in Louisiana. My roommate is keeping them for me."

"Tomorrow is a Saturday and Margaret and Elizabeth can take you shopping," Tom replied. "Susan can wash your clothes, and I will bring you some PJs and a bathrobe for tonight."

Returning in a few minutes with a set of pajamas and a green bathrobe Tom told his house guest, "You can put these on and bring your dirty clothes down when you are ready. Susan is cooking some shrimp, rice, red beans and sausage for supper, and we have some Jamaican Red Top beer to wash it down with. Coming from Louisiana we thought you might like that."

Being something like a new boyfriend being brought home for the first time, but instead being introduced to two young women that he had only briefly glimpsed at the wedding, the supper had the awkward feeling that he was being meticulously scrutinized at every second by eight pair of eyes looking for him to make the slightest mistake.

Step by step he organized his moves. He picked up the salad fork and used that on his salad. He then asked for the sugar, and after tasting the tea, used the long silver iced-tea spoon to add a half-teaspoon to his glass and stir it until it dissolved.

The rice was passed, and he put a mound of it on his plate. Next came the shrimp and beans which he spooned carefully, taking extreme cautions not to overfill the spoon and spill any on the tablecloth or his bathrobe.

The refrain, "Better to be silent and be thought a fool that to speak and remove all doubt," ran through his head. Thankfully Tom spent much of the early part of the meal relating Albert's life story, and while Tom was recounting it he waited for the questions to come from the two sisters.

"What is it like going to college in New Orleans?" Margaret started off.

"It is a bit hectic going into pre-med and working at the same time. Fortunately, Larry Jr. and I shared a room at Tulane while he was working on his E.E. degree. He had some money, and we would go out together when we could. Mostly I worked at bars. He finished up before I did. He went to Atlanta, and I was accepted in med school."

"So, you know how to mix drinks then?" Tom asked.

"Yes, I can put together most common mixed drinks, but I also cooked most of the Cajun bar foods as well. Maybe the most unusual thing that we had for a time were nutria, swamp rat, burgers. There is quite a story about those," Albert volunteered.

Chapter 10

Eats And Meets

"Tʜᴀᴛ sounds disgusting," Elɪᴢᴀʙᴇᴛʜ ɪɴᴛᴇʀᴊᴇᴄᴛᴇᴅ.

"Actually they are excellent, and we had a good demand for them when we could get them," Albert continued. "This is a story about them that takes some time to tell, do you want to hear it?"

Tom made a circular motion in the air for him to go ahead. As he had carried most of the table-talk conversation before it was time for him to eat before his food got cold.

Albert began. "At the orphanage one of the success stories they would tell us was about Billy Joe Rubideoux. He had an alcoholic father, and he was taken away from the family for his own protection and put in the orphanage. He was an excellent student, became a mechanical engineer and graduated from Louisiana State.

"After he got his degree, he wanted to help the people of the lower parishes. With the help of Chinese financing, he set up a factory to process nutria. He would collect the bounty from the state, and sell the hides and meat into the Chinese market. For the first time in his life he made some real money. He married his childhood sweetheart Bobby Jean and they had a son.

"Then disaster struck. Hurricanes Katrina and Rita whipped him out. He stayed at the factory as long and possible and finally took his boat upriver dodging coffins that had been washed out and were floating downstream. One of those contained the body of his mother, although he had no way of knowing it at the time."

"Factory or no factory and hurricanes or not, the Chinese investors wanted their products or their money. There was a little insurance because of where he built his factory, and it was not near enough to cover the loss.

"An uncle who was killed during the hurricane left him a piece of land on higher ground. He built a cook shack there where he made Louisiana specialty products like alligator boudan and sold these along the interstate along with his nutria burgers and other things made from wild hogs and deer meat.

"The state fish and game officials as well as the state health department took an interest in what he was doing, and in particular, where he got this meat. They came to close him down.

"In the meantime, his wife divorced him and demanded alimony to support their child.

"He sold everything and put that money in a flour sack and took his knife and stabbed in into the middle of the table of his cook shack. He took his pirogue, a bottle of Bayou Rum, his mother's bible, his daddy's pistol and polled out into the swamp. They later found the pirogue overturned in a gator hole. People assume that he drank the rum, read the bible, put the pistol to his head, shot himself and ultimately was eaten by alligators. That is the tragedy of Billy Joe and Bobby Jean Rubideoux, or at least that is what he wanted the world to think.

"Later it was said that he was spotted in Portland, Oregon. The last rumors about him was that he was Living with a Mori woman in Goa, India, where they were making money selling very fine glitter powders into the cosmetic industry.

"This is a story of how the Cajuns have often striven to get ahead, and just been generally screwed by everyone and everything, such as the oil companies, the state, rising sea levels and the weather. Even their music has been expropriated and many artists have been left with nothing after a lifetime of hard work. Louisiana is full of such stories.

"Although I am not of that blood, I am of that cultural descent and am fearful that the same thing will happen to me. I may succeed at something, only to lose everything at the last moment. I sometimes wonder, why even bother to try?

"I am sorry to unload on you like that, but that is something I have been needing to get off my chest for a long time. I don't know what made me tell you that. I'm sorry."

"It's obvious Albert," Susan replied. "You have been under a lot of stress for a long time, and your psychological instincts told you that you had a sympathetic audience who would listen, and understand, what you had to say.

"Girls, take Albert to the hot tub and help him relax. Tom and I will join you once we get things cleaned up."

"Before you go, someone else needs attention too," Tom commented.

Rover had been sitting beside the table while the meal was being consumed and when Tom motioned with a piece of sausage came up and gently took it from his fingers. He then worked around the table from person to person and received maybe a bean, or shrimp or another piece of the tasty sausage.

"Rover likes a little spice in his food and enjoys mild chilies, but we try not to overdo it," Tom commented.

Tom handed Albert a two-inch shrimp, "Hold it out for him and let him approach and take it. He likes it tails and all and in particular medium-sized shrimp that he does not have to bite into, like prawns. Don't snatch it back or he will lunge for it and maybe take a finger or two. Let him come up and take it."

Tentatively Albert took the shrimp and holding it by the tail extended it towards Rover. Rover looked up at Albert and then down at the shrimp. Then he took a half step planted his four legs and leaned forward with his entire body as if he wanted to be able to jump back as fast as possible. His nose sniffed the shrimp and then he opened his mouth pushed forward until his teeth touched Albert's fingers and then gently pulled it from his grasp before leaning back to sit and enjoy crunching down his treat.

Tom then held up both hands in a stop motion and said, "No more."

Rover, apparently content that he had received his tribute, went and lay down on the kitchen's tile floor.

"When he really gets to know you he will come up and put his jaw on your leg," Tom informed Albert. "Now you three. Off to the hot tub."

With Margaret holding one hand and Elizabeth the other the two girls pulled a somewhat reluctant Albert up from his seat and towards the bathroom door. As he left he cast back a look of desperation at Tom.

"It will be all right Albert. Just relax and let yourself unwind. You are wound up as tight as a clock," Tom advised.

With somewhat of a note of sympathy in her voice, Susan asked, "Is he going to be all right in there?"

"I think so. I will check on them in a half-hour or so," Tom replied. "In the meantime, I think I'll have another beer. Albert didn't even finish his," he said as he poured Albert's half-drunk bottle into a glass and helped Susan clear the table.

"I need to use the toilet," Albert said as they entered the bathroom.

"They are right over there," Margaret said as she pointed to the toilet stalls which had partition walls between them but no doors.

"You can see," Albert stated the obvious.

"Yes we can," Elizabeth said as she went into one of the stall sat down and urinated into the basin. "You see, like that." She concluded as she took off her shoes.

Albert when into the other stall, undid the bathrobe he was wearing, pulled down the oversize pajamas that Tom had given him and sat down. As he performed his bodily functions, he watched Elizabeth walk over to the shower, turn it on, hang her jeans top and underwear on hooks, shower, soap and climb into the hot tub. Margaret was undressing while he was finishing up.

"Hurry up, the water's fine," Elizabeth shouted.

"Do you want some help undressing," Margaret volunteered as she walked towards Albert. She had taken off everything but her panties, and Albert watched her breast move as she walked.

"No. No!" Albert said emphatically, "You go ahead and shower. Leave the water on, I'll come next." He was grateful that there was only one shower head, so he had a somewhat plausible excuse. He had not gotten to human anatomy were they dissected bodies yet, and this was as close as he had ever come to a naked woman.

Hanging his robe and PJ's up on hooks as they had done, he stepped into the shower and quickly lathered up, and then allowed the water to remove the soap suds from his skin and hair.

He walked over and the two girls were spaced around the edges of the tub. Sitting and sticking his feet into the tub, he said, "This is hot."

"Just ease into it slowly and it will be fine. Let your muscles relax in stages and feel the heat work into your muscles," Margaret said.

"You don't look much like Aron," Elizabeth observed.

"I am sorry, who?" Albert responded.

"Aron, Samantha's boyfriend. Margaret replied. "She is supposed to marry him in May. They came here infested with ticks. Dad and I picked them off him and Mom and Elizabeth worked on Samantha.

"Don't make love in the weeds." Margaret admonished.

Aron blushed as he imagined a teenaged girl picking ticks off his genitals. "They all right?"

"Fine now, I think." Margaret added. "We are all supposed to go up for the wedding in Detroit – nothing as fancy as we had for Beth, but it will be an interesting event. This will be a Jewish ceremony."

"Are you relaxed yet," Elizabeth asked.

"I think so," Albert replied.

"No you are not. You are knotted up tight as a rope," Elizabeth said as she walked over, had him turn around and began to gently manipulate the muscles on his shoulder.

"I think that this will feel good too," Margaret said as she came, sat beside him and lifted one of his legs to place a foot on her thigh. As she rubbed his thigh, she said, "You know you can make love to us if you want."

"I. I never have," Albert sheepishly replied. "I never had any money so I didn't date, and the last thing I wanted to do was to get a girl pregnant before I could support a family."

"You're not gay?" Elizabeth said, "After all you are what 24, 25?"

"No, I haven't had sex with guys either." Albert said.

"I think that we need to do something about that," Elizabeth said. "I have never had sex either, and maybe we should do this together. I know there is nothing romantic between us, but you are a nice-looking guy, and I would like to have you as a first sex partner."

"How are you all doing," Tom asked as he walked in.

Albert thrashed as if to get himself out of a compromising position, but the girls held him in place as their dad approached.

"Albert, this is all right. It is about what I expected might happen," Tom assured the young man.

"Dad," Elizabeth started. "We both want him. How are we going to handle this?"

"As I said, that is up to you three to figure out," Tom affirmed.

"Err. I have never had sex," Albert began hesitantly, "I don't know if I can perform twice in a night?"

"Albert as young as you are, that is likely no problem. It's when you get old like me that we guys get to the 'once and done' stage," Tom replied.

"Albert, since I have had sex before and you have not," Margaret intoned, "I think we should go first tonight, and then tomorrow night you and Elizabeth can sleep together."

"Do what you like," Tom responded, "but that sounds like a practical solution. You know where the bedrooms are, get your clothes together and get out of here, I am going to enjoy a soak."

Tom began to undress while the others got out of the hot tub and toweled themselves dry. Looking like nymphs in a Shakespearian play the three scampered by their startled mother in the kitchen and up the stairs towards the bedrooms.

"Be careful. Don't hurt each other," Susan called to the departing trio.

"Larry, this is Tom. Can you set up a conference call for tomorrow sometimes between us and your dad? Albert, your roommate, is out of the hospital and staying with us for a few days. Tulane is closed and he can't go back to med school. He has gone through pre-med and had a little training in a hospital. He has volunteered to stay with your dad and mom until they can get back on their feet as a live-in nurse. He feels terrible about making everyone sick, and wants to try to make it up to them if he can. Besides, he really has nowhere to go."

"You know Dad and how he feels about strange people in his house. They could use some professional help and physical therapy, although he might fight it like hell if he doesn't like it or thinks it is a waste of time. Does Rover like him?"

"He let him give him a piece of Cajun shrimp," Tom responded.

"Then it might be O.K. Dad always said, 'If a dog doesn't like a man, there is a danger that man is not worth dealing with."

Tom organized the day's activities at the breakfast table the next morning,. Albert was dressed in his clean but old jeans and T-shirt while the two girls were wearing colorful tops and slacks with medium-healed loafers.

I need for you three to go to Wal-Mart and get the basics there. Underwear, T-shirts, socks, three sets of jeans if you can find some that fit. Remember that he is going to have to work in these clothes if he goes to Texas so at least two pair need to fit loosely – something you can jump on a horse with. Then go to someplace like the Suit Warehouse and get him a suit that he can wear to weddings and church and accessories to go along with it. You can pick up a swimsuit and sportswear too.

Albert's eyes began to tear up and Susan asked, "What's the matter?"

"You all have given me the best night of my entire life," Albert sniffled. "These will be the first new clothes that I have ever had. Everything I had at the orphanage was handed down from older children as they outgrew them. Those I bought all came from Good Will, because I had to save every penny. I am sorry," he concluded as he wiped his eyes with his napkin and blew his nose.

"Outside of Larry, you are the first people that I have ever known that gave a damn about me." Albert rose from the table, walked to the sink, grabbed a paper towel off the roller, blew his nose loudly and put the towel in the trash can under the sink.

"I do not know how to thank you all enough," he said when he returned.

"That's quite all right, Albert." Tom continued. "When you get back from shopping Larry has set up and conference call at 3:00 O'clock between us and his dad about your going to Texas and helping him and his wife. You will need to read up on and likely invent some things that they will do that to act as physical therapy, particularly for Larry Sr. He is not likely to waste time doing something unless there is practical benefit from it."

"Albert, do you know what your sizes are?" Susan asked.

"Not really, I had to wear whatever I could put on."

"Elizabeth, write these down," Susan requested as she took a tape measure from a drawer and motioned Albert to come over. Running the tape around his waist she said, 30-inches. After having him standing on a chair she measured his inseam. He felt her fingers touch him as she pulled the tape down to his shoes. Thirty-two inches, she reported. Sitting him down on the chair she measured him around the neck, and called out 12-inches. With a 38-inch chest and a 28-inch sleeve she soon had the measure of the man.

"Now that should get you started." Susan said. "Off you go and remember, get back by three."

"Here goes," Tom said, as he activated the speaker phone. "Everyone is online. Mr. Austerhouse, Albert, Larry Jr.'s roommate at Tulane, is in med school and they have closed the university. He has volunteered to help you and your wife through the recovery process and design a physical therapy program for you."

"Albert. Albert?" Larry Sr. questioned. "Is he the one who ruined my good suit?"

"Mr. Auserhouse, I am sorry," Albert began. "Yes I am the one who threw up all over you at the church. Believe me there was nothing I could do about that, but I can help you and your wife get back on your feet if you will let me."

"I already have help," Larry Sr. responded. "Why should I get another piece of stock to feed and take care of?"

"Dad, he can help you, and the cost of flying a therapist out from Marfa twice a week would be $500 a trip and that would not pay for his charges. Albert will be there seven days a week for as long as you need him. In stock feed prices maybe that is $20 a day for his room and board."

"What does he know about living out here?" Larry Sr. demanded. "Am I going to have to look after him like a six-year-old?"

"Mr. Austerhouse, Tom and your son have told me something of what it is like on your ranch. I have never been out of Louisiana. I know the Cajun people who have had as hard a history and lived as hard a life as you have. The least I can promise you is some good Cajun cooking when I can get the seafood."

"Catfish and fresh-water fish we got in the Rio Grande, and maybe with some canned shrimp and oysters and salt fish you might be able to do something with," Larry Sr. responded. "O.K. you can come for a month, and we will see how things work out. Bring your Cajun spices with you. The only things we are apt to have around here are salt, peppers that we grow and sage."

"Thanks Dad," Larry Jr. replied. "I think that you will like Albert, and he can do you and mom some real good."

"Tom, when do you think that you could have him flown out? Larry Jr. asked.

"Maybe like day after tomorrow," Tom replied. "I can get him by plane to El Paso with no problem and then by train or bus to Alpine."

"Send us a schedule," Larry Sr. replied. "I'll have my hands pick him up in Alpine."

Chapter 11

The Wedding

Moshe Gildersleeve at The Baldwin Reception Hall looked over his appointment book for May 4 as he covered the day's events for the bar manager, "We will have the Goldfarbs today. The options they chose were salvaged funeral flowers, stale beer and pee scent in the bar and spray floral in the wedding hall, day-old from the Kosher bakery and the lowest bid for the ceremony was from Rabbi Newman. They paid for the starter 'Taste of Israel Package' with a spray can of air, water from the Jordan River, a sack of sand, salt from the Dead Sea, a can of olives and a bottle of red wine. The ceremony starts at 10:00 AM. The bar and appetizers open immediately after. Set up for mixed seating for men and women for the wedding and reception.

"This will be regular drink charges plus 15 percent with a cut-off at $400. After that the guests pay. We move the chairs to the edge of the hall after the ceremony and have the Klezmer Band 'Holy Lights' for two hours with the wedding event ending at 2:00 PM. and clean-up after. Everyone in the wedding hall except the horn players and those at the altar must wear masks. Got that?"

"Yea, I got it," Brad Fitzpatrick said.

"And wear the yarmulke I gave you, everybody is Jewish today – at least until 2:00 O'clock."

The 23-year-old bar manager had come from Ireland on a soccer scholarship. With the COVID lockdown he could not fly home or continue his studies in sports business at Detroit University. He had grown up pulling shifts at the family's pub, so it was not difficult for him to fit in as a server, manager or even a cook wherever he might be. He had mostly cooked in Mexican restaurants in the U.S. and prided himself that his Mexican dishes were as good as those made by the other cooks. His Irish accent clashed with the Jewish persona he was supposed to adopt today, but what the heck, he thought, "there are Irish Jews too."

"This room," Moshe instructed, pointing to a small anteroom at the back of the hall, "is to be set with one small table and chairs for two, our best-looking plates, silver and glasses. There is to be a glass each of the wine from Israel, a shaker with the Dead Sea salt, some of those olives, sliced white and dark bread and a plate with sliced meats with lettuce, tomatoes, cucumbers with mustard, mayonnaise and my special glitche sauce on the side. Give them a pasta salad too, and almond cake. Oh, add some water, so they don't load up with too much wine. This will be the first food that they have eaten all day."

"What is this 'Taste of Israel Package' business?" Brad asked.

"I have a cousin in Israel who is making a fortune with this stuff. He gets water from the city system in Jerusalem, compresses the air from the hills above the city, buys Dead Sea salt by the ton, sand from the Sea of Galilee a truckload at the time, packs it up with a bottle of the cheapest wine he can find, puts it in a cardboard box with a cedar print and fancy lettering, and ships them all over the world. He sells air, dirt and water and is making a fortune at it," Moshe informed the doubting bar manager.

"I can maybe see the wine and salt, but what do they do with the air, dirt and water," he inquired.

"They use the Jordan River water like you Catholics use holy water. It is in a sprinkle bottle with a cap on it. Some might be sprinkled on the bride and groom's car, for example. The dirt, they spread out on the steps leading from the church so that they walk on their mother country's earth as they start their new life. They breath the same air as the kings and profits of old to inspire them to think and act big."

"I am not Catholic, by the way," Brad corrected.

"That's fine, then you might as well be Jewish," Moshe asserted.

After reflecting a bit Brad added, "I suppose I see the symbolism in this now that you explain it." Brad responded. "What do these silver letters on the box say?"

"From air, dirt and water the Lord, thy God, made you; and to air, dirt and water you will return. Glory to God," Moshe translated.

"Was that from the Tora or some ancient scholar from Babylon or Alexandria?" Brad questioned.

"No. My cousin made it up two years ago, but it sounds good, doesn't it? Since you are interested, you can help me set up," Moshe said while he walked to a corner of the building where another part of the warehouse had been partitioned off. Unlocking the door, he opened the room. There was an oriental rug on the floor, a chair in the middle of the room and it was lit by a crystal candelabra. The room's interior was covered in paneling. On the back wall was an upright chest with two doors meeting in the middle. Before entering, Moshe put a prayer shawl over his shoulders.

"I need you to help me with the Tora," Moshe said. "It goes on the stand behind the *chuppah* where the bride and groom will stand.

"Help me take it out and carry it," Brad did as he was asked and felt the weight of lidded cylindrical container and whatever it contained. When they reached the stand, Moshe released three brass latches, removed the top and pulled the twin staffs of the Tora from its resting place.

"Help me slowly unroll it until we come to the gold marker where the Rabbi will read."

"How did you get this?" Brad asked. "I thought these were in temples, synagogues or museums."

"I have many cousins. One in New York went to 'Unclaimed Freight' and found this cabinet, the Tora and a shofar for sale. They were in a warehouse that was being torn down. From the best he could find out, it was shipped to New York from Europe just prior to World War II. Maybe the people who sent it thought that they would pick it up when they arrived, but few Jews were being allowed in. I assume they were rounded up and killed by the Nazis or otherwise perished during the war. This Tora was not near as fancy as the ones that pig Goring looted from places like Warsaw, but was from some small-town synagogue where it had been part of that group's history for hundreds of years.

"Each letter was rendered by hand, stroke by stroke on goat hides, sewn together and rolled as you see it now. The letters have a little different look from those I learned in Hebrew school, but because it is a Tora, I was able to read it. Some nights I come down, take it out and study it. How and why I do not know, but I feel kindship with those who over centuries read and heard these words.

"If history teaches us Jews anything it is not to put one's trust in physical things, but to trust in what's here," he concluded as he thumped his chest with the palm of his hand.

Brad noticed that the more Moshe talked the more emotive he became. With his last statements tears began to run down his face and he turned away to compose himself.

"Sorry," Moshe said. "I tear up when I talk about this. If there is any object not touched by the hand of God that is holy, I think it is this Tora. I feel charged with caring for it, and passing it on. This is a living document that speaks to us now as powerfully as it did centuries ago. Something tells me that it should. That is why I bring it out for weddings."

"Sam, are you ready?" Aron inquired. "It is nearly time for us to go."

Wearing a dress with a rose print and lose skirt was out of character of Samantha who felt more comfortable in slacks or the tailored skirt-jacket power outfits that she wore for business meetings. In this event she had to play a more traditional role. Not knowing if Aron's opinion would be of much help or not she asked, "How does this look?"

Aron walked into the bedroom. Samantha stood and slowly turned around.

"It needs a little something," Aron replied.

"What? What?" Samantha demanded. "It is a little late to make changes."

Bringing his hand out from behind his back he produced a box which he opened to reveal a string of cultured pearls which he lifted and attached around Samantha's neck.

"These are from Israel, and I thought it fitting that you wear them on our wedding day," Aron said.

"Oh Aron, these are beautiful and so thoughtful. I love you so much," she asserted as she reached and put her arms around his waist. He bent over and kissed her lightly on the forehead. "Don't want to mess up your lipstick," he said. "We really need to leave."

While Aron was driving them over, Samantha asked, "Who was able to come?"

"I am not at all sure who will be here from my side of the family. There will be Mom, Dad, my sister and her little girl Katy who will act as ring bearer and more or less great aunts, uncles and cousins – some of which I really don't know.

"From your side, Beth, Elizabeth, Margaret, your father Tom, stepfather Fred, their wives Susan and Lollie, and grandparents Robert and Charlotte will be coming. Tyler, your great-step-uncle, I think, was invited, but he is minding their dogs while they are here. Larry Jr. is stuck on a construction job and said he could not make it. I am sorry he can't come. I would have liked to have met him. He asked Beth to send him a video."

"How did things turn out with his dad?" Aron asked Samantha.

"They are out of the hospital in Dallas and back on their ranch in West Texas," Samantha began. "Beth tells me that Albert, the guy who puked on Larry's dad at their wedding, is in medical school and went to the ranch to care for Larry Sr. and his wife until they can get on their feet."

"That will be interesting to hear about," Aron said. "Larry's dad is a crusty old codger who is 'sot in his ways' as Larry Jr. tells it and is not going to take any guff from anyone.

"I think we are getting there."

The streets were lined with old brick warehouses who serviced the auto-parts businesses, food distribution and meat packing. Aron had to slow down because the heavy trucks had pounded deep potholes in the pavement. As a consequence, asphalt and gravel had been poured into the worst of them which made the entire road look like something that had measles.

"Are you sure it's down here?" Samantha asked with a hint of doubt in her voice.

"That's the address. The GPS says that it is just a few buildings down the street."

"There it is. 'The Baldwin Reception Hall,'" Aron said with a tone of relief sounding in his voice. He was beginning to have doubts about the suitability of this place for a wedding, but this was the place his dad had selected.

Pulling up on the concrete ramp marked "loading and unloading only," Aron was met by Brad Fitzpatrick.

"A happy day for you both," Brad said. "Give me your keys and I will have your car parked and brought back after the reception. Anytime you want to leave, ask. I will usually be around the bar."

Brad took the couple to concrete steps cut into the loading dock and helped them up. "We have a freight lift for anyone who can't make it up the stairs."

Entering the hall, the couple saw that folding chairs had been set on either side of a broad aisle in the middle of which was a flower decorated canopied platform.

"You stay here. You will walk down the aisle with your parents. Aron, is it?" Brad inquired. Aron nodded and Brad continued. "You, your mom and dad will go first. Then Samantha you will go and stand beside your husband. You both will sign the marriage contract in front of the Tora and Rabbi. Then Rabbi Newman will read the contract to the audience.

"Samantha, you and Aron will untie this golden colored rope with the knots in it. Then Samantha you will take the rope and walk around Aron three times to signify that you will protect him from all evil laying the rope out as you walk and then rejoin him for the seven blessings. After that, you will both sip from a glass of wine, which at the end of the ceremony will be put in a bag and Aron you will crush with your foot. Samantha you will also step on it to seal the marriage.

"Will any of your family members give some of the blessings?" Bret asked.

"No. We have collected them here. Please let the Rabbi read them. Samantha said as she handed Bret a stack of typing papers printed in a large font. "Aron's Dad wanted to speak on thrift and frugality, but we toned-down his blessing a bit."

While preparations were going on in the back of the hall, the six-piece band was getting set up on the music platform which was dominated by a Baldwin Grand Piano. Ose, the leader of the group asked Moshe, "Has the piano been tuned lately. I like to use it when we do Viennese waltzes after our Klezmer sets."

"The way you play, who could tell?," Moshe rebuffed. "Just get the rest of your instruments up here. Strike the drum and cymbal when you are ready."

The group got their two fiddles, base, clarinet, trombone and drum set ready. With the roll of a drum and the clash of the cymbal the room fell silent and with a single fiddle playing "Sunrise, Sunset" from "Fiddler on the Roof" Aron walked down the aisle followed by David and Sarah.

Sarah, looking at the Rabbi standing behind the canopy and stand holding the Tora, whispered to her husband, "He's so young."

Overhearing David replied, "He's 17, the youngest ever to graduate from Yeshiva University in New York. They won't let him do weddings until he is 22, so he came here. I got him at a discount because he looks so young. He is brilliant, passed all his exams, and is licensed in Michigan."

Samantha followed, accompanied by Tom and Lollie. They looked at the copula and saw that it was decorated in somewhat tired looking calla lilies and branches with white blossoms that were dripping peddles on the floor. "Those look like they came from a funeral," Tom stated.

David responded, "They did. I got them for half off – installed."

Once Aron and Samantha untied the golden rope, Samantha walked around Aron three times while the Rabbi announced for the benefit of the non-Jewish members of the audience, "And thus the wife will protect her husband from all evil and the envious attentions of other women."

When this was completed, Aron brought the corner of the hanging vail and pinned it into her hair, pricking himself in his excitement and his unfamiliarity the process.

As the Rabbi reached for the wine from which the couple would sip after each blessing, Aron grabbed a handkerchief from his pocket to catch the blood.

"There will always be blood," Tom thought. "I hope that this is the worst of it."

Rabbi Newman read the blessing that had been written by David. David blustered up and told the Rabbi, "You didn't mention about how I saved money on the flowers or food or anything."

Holding up the sheet he had been reading from, he mouthed, "It's not here" and continued to the next blessing.

The blessings concluded, the wine glass was emptied, placed in a heavy canvas bag and placed on the floor behind Aron.

Katy, who had been waiting patiently beside her mother, walked forward and presented the rings to the couple who placed them on each-other's fingers while the Rabbi spoke of the continuation of the circle of life represented by this union.

The couple then faced the audience, Aron, thankful that the bleeding had stopped, removed the pin allowing the veil to drop and kissed his new bride. He then turned and forcefully stomped their wedding goblet crushing it beneath is foot and Samantha followed with a much lighter crunch.

"Mazel Tov! Mazel Tov! Mazel Tov!" shouted the mask-wearing audience to the elation of the couple, who were pleased that the event was one more step closer to being over.

Now retiring to the room that Bret had prepared they sat at the table and looked at each other for the first time as husband and wife.

"How long do we have?" Samantha asked.

"At least eight minutes," Aron replied. "We can take a little longer, but not more than fifteen. We must greet our guests and be in the dancing."

"I am hungry," Samantha said.

"We need to eat. There will be food outside, but we are going to be so busy talking and dancing that we really won't have time to eat. Everyone is going to want to talk to us, and we can't refuse."

"The food is good, but this wine is not so hot," Samantha said.

"It is from Israel," Aron said as he examined the bottle. "That is about the best that can be said of it. There will be better wine outside, I think, but be cautious. We are going to be tossed around on chairs like bags of beans. It's meant to be very lively and exciting."

"It sounds scary and dangerous to me," Samantha said.

"Just hold on tight and wave your handkerchief when the men bring us together and that will end it. there will be other dances, but you don't have to participate if you don't want to. Just watch us men make fools of ourselves. There will be some waltzes later when you can dance with your dad, my dad and me. Then we can leave the hall, have more pictures and go home."

"I'll be glad when this is over," Samantha said.

"We could have more days of smaller parties with different family members, but with this COVID business, I don't think so. We are going to wear our masks and everyone else is too unless they are eating or drinking at the bar," Aron concluded.

After taking as much of their allotted fifteen minutes as their dared, the pair emerged to another round of Mazel Tovs. The chairs had been cleared away and the Holy Lights struck up "Halva Nigella." A dozen young men now dressed in black pants, white shirts with their shirttails hanging over their trousers and black vests were dancing in a circle as four men approached Samantha and Aron with chairs. After sitting Samantha down two men picked up the chair and put it on their shoulders. The larger Aron required four men with one holding each leg of the chair. Once seated and up they joined a circling ring of dancing men while the newlyweds hung on.

Samantha freed one hand and gamely waved a white handkerchief as if teasing Aron to come and take it. Ultimately, he was brought close enough to grab it, symbolizing the completion of their union. With a round of applause and more Mazel Tovs that stage of the event was ended.

They later circulated among their guests and with the waltz music signaling the approaching end of the event, Samantha danced with her father, stepfather, Aron's father and finally with Aron.

Walking up to the pair Brad said, "All of your obligations have been completed, do you want me to have your car brought around?"

Getting looks of "please get us out of here" from them both, Brad called for the car. Gathering the spray can of air and sprinklers of soil and water from Israel he escorted the couple from the dance floor onto the loading dock. With Rabbi Newman looking curiously at the proceedings, he sprinkled the soil on the stairs and as they approached sprayed the air with the can marked with the Star of David. When the car arrived he sprinkled the water from the Jordan River on its hood and gave them a parting shout of "Next year in Jerusalem." He felt satisfied that his Jewish obligations, whatever they might have been, had been discharged.

Chapter 12

Conception

Once again, Samantha was in Dr. Mayfield's antiseptic office with Aron sitting nervously beside her twisting a paper wrapper from a piece of peppermint.

"Aron," Samantha said. "It will be all right. You don't have to do anything this time."

"Hello." Nurse Elrod addressed them both. "This is your big day. The doctor is ready for you. Mr. Goldfarb. The hospital likes husbands to observe, and would like to go over a few things with you first. Don't worry. We are going to put you in a wheel chair this time."

Aron, blushing with the memories of his nearly knocking himself out when he tried to give sperm and passing out when the doctor came at him with a needle, replied with a hesitant, "I suppose so. It is going to feel strange watching someone else have sex with Sam, whatever the setting might be."

Nurse Elrod motions and Aron reluctantly left his chair and followed the nurse into the doctor's office.

"Mr. Goldfarb it is good to see you again," Dr. Mayfield said as he rose and extended his hand across the desk. "While Nurse Elrod is getting Samantha ready, I wanted to tell you what you can expect to happen after the implantation and during pregnancy. You are a young couple, and I assume that you and Samantha have, and want to continue to have, a healthy sex life."

"We really haven't since we learned about this chromosome thing," Aron replied. "I want to. I want to bad; but we did not dare, even with condoms. Knowing that we were going to do this … ur, procedure, we didn't want to take the chance that we would have a child with an untreatable disease. We both wanted it, but neither of us would."

"I wish all of our patients exercised such self-control." Dr. Mayfield commented. "After implantation I want you to continue to abstain from sex for at least a week. We want to make sure that the egg is well bonded to the uterus. At six weeks we will be able to check on its development. By that time, your wife will show signs of being pregnant. Then, with some adjustments, sex is usually no problem except during the last trimester. Your pediatrician will be able to guide you through these stages. Generally, if it hurts or anything looks or feels strange to her, quit."

"You said adjustments?" Aron inquired somewhat squeamishly.

"Once her abdomen expands it's uncomfortable for the woman to have face-to-face sex," Dr. Mayfield continued. "Her kneeling, or on top or you both on your sides will often be more comfortable. Because of increased blood flow and hormones, women will often have the highest sex drive and best organisms of their entire lives during this period. It is helpful if you can be ready for this, and can accommodate her desires."

"Do you have any questions?"

Aron thought that, "Yes he did – a thousand of them," but as he could not think of a single thing to say at the moment. His reply was a weakly expressed, "No."

"Very well then." Dr. Mayfield said as he rose. "I will get scrubbed up and then we may begin. I will come for you when we are ready."

"Honey-child," Nurse Elrod said as she helped Samantha on the exam table and prepared her for the procedure. "You are going to have a fine baby, and I don't want you to worry about it. In fact, I have a secret to share with you."

"What?" Samantha asked.

"That man of yours has got some fine stuff, and you are about to have the best fucks of your life while you are pregnant," Nurse Elrod asserted. "You tell him to shave up and be ready. They don't like it when I take that hair off. It is like cutting the long feathers off a fine bird dog's tail. They feel embarrassed and run around with their tails between their legs. But you probably don't know about that? Some people down there in the South have fine hunting dogs, setters and Spaniels, that they think more of than their wives. Sometimes with good reason, let me tell you.

"During that second trimester you are going to feel so hot and so ready that you are going to want to ride that man like a pump handle – up-down, up-down, up-down until he is dry. Just go get him. Towards the end your organisms might hurt the baby by starting premature contractions, but until then go after him like you were a bear trap."

"Samantha, Aron" Mayfield addressed when he had them both in the treatment room, "this will be the reverse of the procedure I used to extract the eggs. This time I will implant the fertilized egg that is in this container," he said as he pointed to a cryogenic container that had a coating of smoking ice on the outside. "This is time critical. Once the egg is ready, I have only a few minutes to insert it into the womb."

Seeing that the couple apparently had adsorbed this basic information he continued. "It is possible to implant multiple eggs. In the early days of artificial we did which resulted in multiple births. In your case that is not desirable, but it is a possibility. I need for you both to sign these forms acknowledging these, and other, risks related to childbirth before we proceed. Do you have any questions?"

Samantha took the multi-page form, turned to the rear signature page and signed and passed the clipboard to Aron. Aron took the documents hesitantly and saw words like "miscarriage, stillbirth, malformations, bleeding and prolonged symptoms and he hesitated not knowing if he wanted to put the woman he loved and this potential child that was about to be "made" at risk.

"I know you want to read every word, dot every 'I' and cross every 't,'" Samantha said comfortingly, "but just sign it. We both want this child and we have gone through all of this before. Just sign it," Samantha requested.

Aron, gritted his teeth, took the pin, turned to the last page of the document and signed.

"Very good," Dr. Mayfield said. "Aron you sit in that wheel chair against the wall. You will be able to see the screen that I use to guide the needle to plant the egg. Nurse Elrod is going to put a belt on you just in case you faint. Samantha, I am going to give you a local anesthetic, like I did before. Following that I will extract the egg from the cryogenic container, check it and then do the implantation. One last time, are both of you sure that this is what you want to do?"

While Samantha replied with a definitive "yes," the best Aron could muster was a nod of his head.

After Samantha replied that she was feeling numb, Dr. Mayfield began, and the pageantry of his tool's journey through Samantha's body began. Aron imagined that he was in a small boat traveling through a damp pink cave as he watched. Like a river boatman, Dr. Mayfield knew which turn to take and where to push the needle through to place the egg in its optimum position. Taking no time to admire his handiwork, he withdrew the probe and the procedure was over.

"That's it," Mayfield said. "Come back in six weeks and we will make sure everything is all right from the genetic and implantation side, and then you will have regular check-ups with your pediatrician. Samantha you probably won't start to see or feel anything until about the end of the first month. Then you might have morning sickness and things like that. If you have pain, start bleeding or have unusual smelly discharges immediately come to the emergency room. Post-implantation symptoms are rare, but they do occur.

"Aron, how are you doing?"

Somewhat sheepishly, Aron replied. "I'm fine. I guess it happened so fast that I did not have time to react."

Driving back to their apartment, Aron turned to Samantha and asked, "How are you feeling?"

"Now that the local is wearing off, a little strange down there," Samantha said, "but I don't feel any pain or flow or anything."

"I really got the 'fuck talk' from Nurse Elrod, did you get the same from Dr. Mayfield?"

"I guess," Aron replied. "In school they didn't tell us hardly anything. We were just supposed to know. Our teachers apparently figured it out, and somehow we were supposed to do the same. Hebrew school was not any better. We learned all the begets, but nothing about the begetting.

"What we learned about women was that you were supposed to be dutiful, obedient wives; but they would always follow that up with some story about how Dalia tricked Samson and cut his hair, or some other tale about a woman tricking someone into doing something – sneaky you were."

"We do have our wilds," Samantha coyly said as she looked up at Aron almost distracting him from his driving. "Nurse Elrod said that during the second trimester that I was really going to want it and to 'ride you like a pump handle' until I had enough. She promised me that I was going to have the best fucks in my life."

"She used that word?" Aron questioned

"She sure did," Samantha enthusiastically replied. "She also said that we should really stock up in condoms, particularly the purple knobby ones, because they might be hard to get with everyone locked up like we are."

"She talked about me too?" Aron questioned.

"Yes she did. She said for you to keep staying shaved and give me what I wanted or send you back to her and she would get out that razor and do it again."

"She said that?" Aron asked with a note of disbelief.

"She said that you were a fine-looking man and to get out the velvet ropes and whips if I needed to keep you in line," Samantha informed her husband.

"But we don't own any," Aron replied.

"We will, husband dear, we will," Samantha concluded.

Chapter 13

Aftermath

"This apartment was fine for just us, but we are going to need a bigger place when the baby comes," Samantha said. "I want a place where we can have a nursery for the baby and a bedroom that someone can use if they come for a visit. Mom said that she would come for a week when I have the baby and she needs to stay here. It is also going to need an elevator or be on the ground floor."

"And we are both going to need somewhere to work," Aron added. "That means at least a three-bedroom apartment if not something larger. We've got nine months to sell my apartment, find another and move. I would really like something closer to the hospital."

"There is no reason to panic about it." Samantha said as she got out a yellow pad and started to draw action lines down the length of the paper and date lines across. "If we plan step by step, we will be able to get all of the financial stuff done, a new house purchased, and out stuff packed and moved before the baby arrives.

"Your sister, Rose, she's in real estate, isn't she?" Samantha queried.

"Yes, and there is no reason that she can't help us like any other client," Aron affirmed. "We need to know how much we can get for this apartment so we know what will be reasonable for us to buy. It has been quite a relief not to pay rent, and I want to put as much down on it as we can. We are both making reasonable money, but this housing market is going out of sight, and the safe thing would be to plan on making payments out of just one of our salaries if we have to."

About three months down on her time-line Samantha wrote "price house and arrange for sale."

Samantha told Aron, "If we can get a price set and contract on our place during the first three months, that will give us a feeling as to what kind of money we have to put down on something new."

"I cannot say for sure," Aron suggested, "but Joe, Joe Prichard, downstairs has told me that if I ever wanted to sell this apartment that he was interested in it. I don't know what the situation is with Joe and his job, but let's invite them up and see if they are still interested. It that works, it would be great."

"Joe, Kath, I am glad you could come up," Aron said as he greeted them at the door. "Please sit and I will fix us some drinks."

"I'll take a beer," Joe said.

"Some vodka and orange juice for me," Kath responded.

As Aron went to get the drinks Samantha came in and sat in a chair opposite the sofa where the Prichards were sitting.

"Sam, how did that implant go?" she asked. "Did it well, uh, 'take' for want of a better word?"

"It is still early, but I don't feel signs of anything strange, so I suppose, I hope, that everything is fine. The doctor said that the embryo is so small that many don't even know if they are pregnant for six weeks. We are sitting on pins and needles waiting to find out, but the doctor said, 'Just be patient. All will be revealed in good time.'"

Aron returned with the drinks. "Sam, I brought you a plain orange juice on crushed ice, a Miller Lite for Joe and for Kath, a vodka-orange. We are starting to plan for when the baby comes. This housing market is going crazy. I talked to my sister, Rose, who is an agent, and she says people are often bidding above ask price just to get a place. Fortunately, some are selling out in the city and moving to where they can buy a few acres and have more room since everyone is at home all the time."

"Aron, Kath and I are not going anywhere," Joe said. "We like the neighborhood, and this building is close to our jobs and everything we need. Like I told you we would be interested in buying this apartment, rather than paying rent as we are now. Then we would wind up owning something."

"That would make your moving easier," Samantha observed. "All you would have to do is move up one floor."

"I wish we had teleportation," like in 'Star Trek,'" Joe mused, "but just moving up two flights of stairs beats hiring movers to take us across town."

After they finished their drinks, the Prichards asked if they could look around. While Samantha escorted them, Aron fired up a small hibachi on the porch where he was going to grill some salmon, peppers and eggplant for his guests.

With a choice of ice tea, beer or soft drinks, the four sat at the kitchen table to enjoy their meal. He cut the salmon into two-inch wide strips and encouraged Joe and Kath to serve themselves while he washed up and returned to the table.

"This is really good," Kath remarked as she sampled the dishes.

"I put a little coarse salt and pepper on the eggplant and some Parmesan cheese on it," Aron said. "Nothing at all fancy, but it beats paying fifty bucks for it in a restaurant."

"About the apartment," Joe said. "We like the layout of this one better than ours. Ask your sister to come up with a market price. If we can live with that, and I suspect that we can, we will go ahead and get all the paperwork done in time for you to move into your new place. That way you will be all settled in before the baby comes."

"That would be wonderful," Samantha replied.

Although it seemed that six weeks took very much longer to pass than usual, once again the couple was in Dr. Mayfield's office waiting for the first ultrasound of their developing baby.

"Samantha," Nurse Elrod said. "Come in and I will get you ready for the exam.

"Has he been treating you right?" She said as she pointed at Aron, who blushed at this rather direct question.

"I haven't had to get out the whip yet," Samantha replied with a kidding tone to her voice.

With a parting, "You better do right by this little lady, or I will come after you with my razor," Nurse Elrod quipped, which elicited an involuntary shudder through Aron's body as he thought of once more going under the knife administered by the not quite so tinder hands of Nurse Elrod.

Once Samantha was prepared, Dr. Mayfield placed the probes on her skin and focused them to deliver the best image on the screen. Once he had it centered, he hit the record button and as a video started to be made the tiny fetus could be seen responding to every beat of Samantha's heart.

"It is early yet to tell anything definitive about the fetus," Dr. Mayfield began. "At this stage it looks absolutely normal. I am going to take a blood sample from your vein exiting the placenta and have that tested for the chromosome defect. I will have that result tomorrow. If all of that looks good, I will pass you onto Dr. Richard Kitchens, who is your assigned pediatrician. He will follow your progress through fetal development and childbirth."

"We won't see you again?" Samantha asked.

"Not unless some problems develop," Mayfield responded. "I will call you tomorrow and let you know about the test. There is no reason for you to wait weeks to find out. We are trying to keep traffic in and out of the hospital to a minimum."

As the pair left the doctor's office, Nurse Elrod walked up beside Aron and said pointing her finger at his groin saying, "You better do right."

"Don't worry. He will," Aron responded. "I am sure he will do fine."

The next day Samantha was on a conference call and Aron was talking to a couple about their insurance issues when the apartment telephone rang.

"I'll get it," Samantha said.

"Gentlemen," Samantha said to those on the conference call. "I need to take another call. Please continue. I will rejoin you in a few minutes."

Picking up the telephone, she answered, "Samantha Goldfarb here."

"This is Dr. Mayfield. I am pleased to say that your chromosome test came back fine. I have set you and Aron up with an appointment with Dr. Kitchens in six weeks. Nurse Elrod will send you both e-mails. If you need to change this appointment you may, but appointments are being restricted so please make it if at all possible. It is important that the husband also attend."

"Got it thanks," Samantha replied.

Putting down the phone, Samantha took off her headset and ran into the room where Aron was working. When Aron turned his chair to face his approaching visitor, Samantha took the headset off his head, kissed him, and said, "That was Dr. Mayfield. The tests are back, and the baby is fine." Not waiting for the startled Aron to reply, she replaced his headset and ran back into her room to continue her call.

Aron took a few minutes to mentally detach from the life-changing news he had just received and returned to listen to the couple on the phone describe their difficulties in getting responses from insurance companies. It was hard to reengage as his mind realized that he was going to be a father and all of the events that might result. His physic worked best with things things he could control, and he was to be experiencing a universe of events that were unpredictable and unknowable.

"David, pick up the phone," Sarah shouted to her husband from the kitchen. "It's Samantha. They have received some word about the baby."

"Yes, Samantha, we have all been waiting to hear," David replied. "How did the tests come out?"

"They took some of my blood," she reported, "and it came back negative. Although the fetus is very small at this stage, the doctor said that it was developing normally."

"Thanks be to God," Sarah interrupted. "With all that you and Aron were going through, I am so very happy and pleased."

"Aron and I appreciate living in the apartment that you bought for Aron, but with a baby it is going to be too small," Samantha said.

"Rose is going to help us find another. We plan to sell this one and pay down as much as we can on something that is larger, on the ground floor or has an elevator."

"So, what I gave him wasn't good enough?" David rebuffed.

"It did just fine for a young single man," Samantha explained, "but now that we are two and soon to be three, it is not big enough. I also don't want to risk carrying a baby and stroller down four flights of stairs."

"David, she does have a point," Sarah interjected. "With a baby there is going to be heavy cases of baby formula, a stroller to deal with and bulky boxes of diapers and who knows what else to get to the apartment."

"My mom did all that, and carried coal and water too," David rebuffed.

"She was a much larger person than Samantha," Sarah said, "And she complained of back pain all of her life. Had she been spared some of that, I am sure her last years would have been much more comfortable."

"Maybe so," David relented. "We want to get you something for your baby. I'll look in the paper for unused baby items that a couple might have bought who lost their child and are selling off the baby furniture that they bought."

"You will do no such thing," Sarah said. "I will not have our grandchild jinxed because he is sleeping in some dead baby's crib.

"Samantha, we will get you some things for the baby's room, and they are going to be brand new," Sarah emphasized.

"I don't know if you have any idea of if it is a boy or girl, do you?" Sarah said.

"No." Samantha replied. "We don't, and we don't want to. I want that to be a surprise."

"We can buy things for a grandson and then you can use those until Aron produces one. If he is able," David said.

"Mr. Goldfarb," Samantha rebuffed, "Let me tell you that your son is plenty 'able' in that department. We will be needing a crib. That can be something gender neutral with baby-safe paint. If you want to give us something for the baby, that would be appropriate."

"Thank you Samantha, for letting us know, Sarah concluded. "If you need us, call."

"Your turn," Samantha told Aron. "I called your parents, now you call mine."

Taking up the phone Aron punched in the numbers and Tom answered. His answering rather than Susan made Aron somewhat more comfortable.

"Tom this is Aron," he began. "We got the results from the blood test earlier today, and the baby is fine. It's still too small to tell anything about, but the doctor said that on the ultrasound, it was moving and acting as would be expected."

"That is wonderful news," Tom replied. "I don't imagine that you want me up there for anything, but at some stage Susan, Beth, Elizabeth and Margaret might come up alone or maybe together for a baby shower or something. I will let them know. Just keep us posted. We are all pulling for you and Samantha down here."

"How are Beth and Larry doing since they married?" Aron asked.

"Both are working like blazes. Larry is literally up to his neck in concrete and wires, and doesn't know when he will ever get back, and Beth is looking forward to school closing and getting some time off. She might even fly to South Dakota to spend a few days with Larry. Thanks for the call."

"See that wasn't so bad," Samantha said. "Just one more for now. You need to call my mom. I'll dial it for you."

Aron took the ringing phone and was mentally preparing a voice-mail message when Lollie picked up.

"Mrs. Williams this is Aron. We got the results from the blood test today and the baby is fine. It is not very big, but the doctor assured us that it was acting normally. We are expecting a delivery date in November or early December."

"They don't know better than that?" Lollie responded incredulously. "Nine months after the date of conception, the math is not that hard."

"That is true, but because of Samantha's small size and the possibility that they may have to take the baby early, he did not want to be definitive."

"I want to come up and be with Sam for the birth and stay for that first week," Lollie responded. "I will need to take off time from work and arrange for it in advance. I can let everyone know and keep it a bit lose for now, but as close as we come to a delivery date, let me know."

"We are also going to be buying and moving into another house," Aron added. "We have a buyer for my apartment, and my sister is going to help us find a bigger place so we can have a baby room, office, a spare bedroom and an elevator. We already have most of our furniture and need only a few things."

"Fred has a saying that, 'Shit expands to fill all available space,' so make sure that you get a big-enough house," Lollie advised. "With that new baby, you are going to fill it up."

"Thanks, Mrs. Williams, we will certainly keep that in mind," Aron said as he looked at Samantha to see if there was anything else that needed to be said.

With a shake of her head Samantha signaled "No," and Aron ended the conversation.

A few days later Aron and Samantha were again working in their apartment when the house telephone rang. Samantha was between calls and picked up the phone.

"Samantha, this is Rose, I think I have found you a place."

"Aron, pick up. It's Rose. She thinks that she has found something for us."

"I have a listing that just came on-line that we need to look at today. This is in a new subdivision called Springdale. The apartment is about two years old. The owners inherited a property out of town and want to move there. It has three bedrooms, two full baths and a large open room that is a combination great room, kitchen and dining room. It is on the third floor, but has an elevator.

"With what I negotiated with Prichard for your apartment, you can get this one with total payments of about 85,000 over ten years at just under four percent. I talked to the owners and told them your story. They feel like they have received 'money from heaven' and were shocked to find out what their apartment was worth today. They are willing to take their asked price from you, without inspection and without improvements if we can settle quickly. We have to move on this one because people are willing to pay ten percent over their price just to get a half-decent place in this neighborhood. I told them we would look at it this evening. Can you come?"

Very quickly the pair agreed to meet Rose that evening.

"Fortune does not favor indecision," Samantha quoted from one of her business books. "If this place looks reasonable, we should grab it. We may have two apartments to clean up rather than one, but we have the time."

When Samantha, Aron and Rose visited the apartment they were equally shocked and surprised at what they saw. Its location on Springdale was closer to the University hospital as Aron wanted. The front of the two-story set of apartments consisted of a large garage door for three vehicles in the basement with a narrow walk leading to a door which allowed entry into a stairwell. The ground floor was almost all garage with a laundry room, three small caged-off areas for storage of items like snow tires and an elevator.

The apartments were located on the first and second floors. Each of them had two large windows open to the front of the street and smaller ones along the sides and rear of the building. There was also an external fire escape at the rear of the house which discharged into a small fenced-in yard.

"Modern utilitarian," Rose remarked. "Nothing at all fancy – just a working home for working people."

Entering the second-floor apartment Aron and Rose found scattered paper, box parts and tape fragments littering the floor and nails were on the walls where pictures once hung. One broken chair and a very much abused sofa remained. Notably missing were the stove and refrigerator with only plugs and outlines on the tile floor marking where they had once been.

"I don't know for sure," but I think our appliances might fit.

Rose, taking a tape measure from her bag had Aron assist him in making the measurements. "Now we'll know. There is no reason to move them if you can't use them," she quipped.

"I'd say the place has potential," Samantha said. "It is not quite as big as we wanted and doesn't look as nice as I would like to have it, but it will do for now."

"Are you sure Sam?" Aron queried hesitantly. "This place is going to take a lot of work, and there are just the two of us."

"What we can't do we can hire done if we start soon enough. Everyone seems to be remodeling right now, but we still have seven months to get everything hooked up and going. Even for Detroit, that should be possible."

"I will start the paperwork process and hope to get everything signed in a week or so. Then the utilities can be turned on, and you can start moving in."

Chapter 14

Unexpected Events

SITTING NERVOUSLY ON THE EXAMINATION table dressed in expanding green slacks with an oversized orange pullover restraining her bulging belly, Samantha waited for Dr. Richard Kitchens. The 45-year-old physician came in dressed in a white lab coat with a brown checkered shirt and tie. After sitting at a desk, he scrolled through several pages on his computer screen. Looking over the rims of his glasses at his young patient, he hesitated for a moment.

"Mrs. Goldfarb, this has been a risky pregnancy because of your genetics predisposing the baby to liver disease. The good news is that the fertilization, testing and implantation worked, and the baby seems to have developed normally."

"What's wrong," Samantha asked.

"Your blood pressure is high, and the meds aren't working. I think that the prudent thing to do is to induce labor and deliver the baby today."

"Today," Samantha interjected. "I'm not ready. Will the baby be all right?"

"Although the baby will be a month early, it looks to weigh about five pounds. Everything should be fine. Is your husband in the hospital today?"

"No. He's working at home. I need to call him and have him bring my bags."

"While I arrange things here at the hospital you can call whoever you need to notify," Dr. Kitchens suggested. "It will probably be a half-hour before things get started."

Aron was sitting at his desk in his apartment. Even though he was working at home he was immaculately dressed, clean shaven and wearing a tucked-in starched shirt with a pocket protector with three exactly spaced pins and Chicago Cubs tie. Surrounding him were open boxes, newspapers and strapping tape. Startled by the ringing house phone, he spilled his Trader Joe's coffee on his pale bare legs and wiped it off before it could stain his fluffy white bedroom shoes. He picked up on the third ring.

"Aron Goldfarb. University Hospital."

"Bring my bags to the hospital," Samantha said hurriedly. "They are going to deliver the baby today."

"Sam, Are you O.K.? That was not supposed to happen this soon. Is the baby all right?"

"It's fine. I'm fine. Everybody's fine. Just get here. Bring my hospital bags and your stuff too."

"I'm not packed," Aron protested. "I'm working with a couple, and it is snowing."

"I don't care. Get your ass here and call Mom. And bring my bags!"

"You don't have them with you?"

"No I don't because I didn't know I was going to have the baby today. Get your shit together and get over here, and call my mom."

"I had rather not talk to your folks. I'll call mine, and you call yours."

"Whatever. You just get here."

Samantha dialed her mother's number. The call found Lollie in the kitchen and Fred sitting in his recliner watching a NASCAR race. Fred picked up the ringing phone.

"Hello. Hum. Is that so?," Fred responded and then called to his wife, "Sam is going to have her baby today."

Lollie wiped her hands and picks up the nearest phone, "Sam. Today? Did I hear that right?" she asked excitedly.

"Yes. My blood pressure is high, and they want to birth me today."

"There is a bad snowstorm coming to Atlanta. We will be there as soon as we can," she concluded.

Packing more rapidly than either of them thought possible, Fred and Lollie assembled what they thought would be their warmest clothes for their trip to the North Country. Fred's heaviest hunting outfits were too bulky, and he settled for a medium-weight camo top and bottom and his cleated hunting boots. Lollie selected some wool suits and a topcoat along with scarves and a hat that was more functional than decorative.

The edge of the snowstorm had arrived as they approached the airport. They had phoned in their ticket request to Delta and had their tickets on their phones. Fred would have preferred paper tickets; but was nonetheless ready, given the circumstances, to enter the modern world.

Whatever they carried and whatever they wore were going into the cabin with them. It was with considerable relief they were passed through the gate and ran up the gangway to find a place for their bags relatively near their seats. Because this was a last-minute booking, they found that although they were on the same side of the aircraft Lollie was two rows back of Fred.

"Excuse me sir," Fred said as he addressed a suit-wearing black man in the aisle seat. "My wife is two rows back. We are trying to get to Detroit where our daughter is having our first grandchild. Do you mind changing seats?"

The black man, who looked and sounded something like Rochester from the Jack Benny show, looked back at the seat where Lollie waved at him.

"Doo tell. I wasn't there for any of mine. One was born in New Orleans and the other in Japan. Why I remember….."

His response was interrupted by the pilot, "This is your Captain speaking. Because of icing our taking off depends on how fast you can get into your seats. There will be no cart service and no moving around once the aircraft takes off.

"Prepare the cabin for departure."

"I'm sorry, but we don't have time to talk," Fred pleaded. "Will you change seats?"

"Show will," the seatmate replied. "Glad to help. Best of luck with everything. God bless."

Lollie rapidly moved up and thanked the man as she passed. As she sat down, the huge handbag she was carrying spilled open, and she and Fred picked up the cosmetics and other things she was carrying and hurriedly stuffed them back into the bag. A passing steward helped, and soon Fred and Lollie had the bag crammed under the seat.

"I don't see why you have to carry so much stuff," Fred remarked.

"You never know what you might need," Lollie replied. "You never know."

With their hands tightly gripping each other during the take-off run, they looked at each other with a mutual sigh of relief in acknowledgement that they were on their way – storm or no storm.

As David entered the kitchen of their Detroit house from the mud room with a bag of groceries, he was dressed in a skill cap beneath which a ponytail of white hair was held by a rubber band. When he came through the door, he kicked off his rubber flip-flops and stood in his stocking feet.

"That was Aron's wife on the phone." Sarah began. "She is going to have her baby today at University hospital."

"I thought that would be next month," David replied with a note of concern. "Is she and the baby O.K.?"

"Samantha said that she's fine, and there is nothing wrong with the baby. Her blood pressure is high and they are going to induce labor."

"What do you want to do?" David asked.

"I want to go to the hospital. Someone should be there."

"They were supposed to move to their new house before the baby came," David ventured. "I know they have been cleaning the new one up, but they are not near ready."

Back at the apartment Aron retrieved an old Samsonite suitcase and positioned it on his bed so that it exactly lined up with the corners.

Returning to his closet he selected one of four sets of identical gray slacks, pulled them on over his Chicago Cubs boxer shorts, then completed his over-meticulous dressing as if he was to appear before royalty.

Samantha was on a gurney covered in blankets while around her was a bustle of activity as she was prepared for delivery.

"Is my husband here?" Samantha asked.

"I am sure he is on his way," Dr. Kitchens replied. "His parents are here. Do you want to see them before we get started?"

"No." Samantha replied definitively. "If we are going to do this thing, let's do it."

"Next, we'll take you into the delivery room. I'll start the drug that will induce labor. This drug may take some time before you feel anything. When you want anesthetic, tell me."

David and Sarah are sitting side-by-side on a couch when Dr. Kitchens, now masked and dressed in scrubs, walks in. "I've started to induce labor. We have to be very careful about the drug dose, and I do not know how long the process will take. Where is the husband?" he asked as he looked around.

"I don't know," David replied. "I suppose that he is on his way. I'll find out."

Aron, now dressed, was carefully layering his shorts, undershirts, shirts, slacks, socks and toiletries in his suitcase when the bedroom phone rings.

"Aron, where are you?" David demanded. "Your wife is starting to have your baby."

"I'm home. Packing."

"Get over here now!" David ordered. "Your wife needs you. Get over here or she will never forgive you."

"It is snowing." Aron replied.

"I don't care if it is raining fire and brimstone," David responded, with an increasing air of exasperation. "Leave now! That baby is not going to wait on you."

"Don't you think that you were a little hard on Aron?" Sarah asked.

"That boy frustrates me so bad, that I sometimes feel that I could strangle him." David responded trying to calm himself. Getting control of his emotions, he thought it was time for him to change the subject. "How do you think the delivery will go?" he asked his wife.

"Samantha is so small that I cannot think that it will be easy," Sarah replied. "That is going to be a big child. I remember when my cousin had her first the labor went on for hours and hours and the baby did not survive. With everything else, she could lose the child."

"Don't think like that," David said in as comforting a voice as he could muster. "This is one of the best hospitals in the country and they look after their own. I'm sure everything will be all right."

Aron, breathless from running across the parking lot and walking as fast as he could through the hospital, burst through the door of the waiting room.

"Is the baby here yet?" he asks.

"No. But no thanks to you. What took you so long?" David asked.

"The snow and traffic were terrible," Aron replied.

A male orderly walked in and seeing Aron asked, "You the father? I need to get you ready."

"What's happening?" David asked. "It's been two hours."

"The doctor will speak to you in a moment," the orderly replied.

Dr. Kitchens, entered in a hurry and informed the family, "Samantha is fine, but the baby is breached, and we are going to have to do a C-section." When he completed his statement, he turned to the orderly and said, "Get him ready. As soon as I scrub, we will began."

Aron now wearing a new surgical mask and dressed in scrubs and rubber gloves stood beside the delivery table. He took Samantha's hand and gently held it.

"Sam. I am here," Aron announced.

"About damn time. I thought that I was going to have to do this alone. Give me something now, and get this thing out of me."

With this statement, the anesthesiologist put a mask over Samantha's face and instructed, "Take a deep breath. You won't feel a thing."

After the anesthesiologist checked Samantha's breathing and heart rate, he nodded to Dr. Kitchens who removed a scalpel from a tray and started his incision. Aron swooned and was caught by the orderly and laid down on the floor. The orderly drug him away from the operating table and propped him up against the wall.

Aron revived to the sound of his baby's squall, shook his head and blinked his eyes as they readjusted to his surroundings. The orderly came over and looked him in the eyes.

"Are you all right?" the orderly asked.

"I think so. What happened?" Aron inquired.

"You have a son, and your wife is fine," the orderly responded. "You can see them later. Can you stand?"

"I think that I am all right," Aron responded hesitantly.

"Sit for a few minutes. I've had guys faint two or three times before they get out of the delivery room. A smell, a look, anything, can provoke a response," he informed the new dad.

"Don't tell anybody, will you?" Aron pleaded knowing full well what his dad would have to say.

"Never will. Never do," the orderly replied. "You seem to have recovered. I'll walk you outside."

As Aron entered the waiting room he announced, "I have a son, and Sam is okay."

He was immediately followed by Dr. Kitchens who informed the new father and his parents, "Samantha came through very well, and the baby is fine. We will have a report from the chromosome study shortly. Samantha has been taken to her room, and you will be able to see her in a few minutes."

"This is a blessed day on our house," Sarah commented. "I am so very happy and relieved."

David searched in the pocket of an overcoat that he had draped over a chair and pulled out a foil-wrapped cigar with a blue ribbon tied on it. "No one smokes these anymore. I gave these out when you were born. This is in honor of your new son, and I have three more," he concluded.

Six hours later, Fred followed closely by Lollie quietly opened the door to Samantha's room and entered. They found Aron sitting in one chair, his mother nodding off in another and Samantha asleep in the bed with monitors and IV lines connected to her. When Aron saw the couple, he got up and motioned for them to step inside the hall.

"I am so very glad that you could come," Aron whispered. "Sam had the baby last night about seven. It is a five-pound boy and both are good. They want to send us home as soon as possible because of this COVID-19 stuff - maybe day after tomorrow."

"When can we see the baby?" Lollie asked.

"The baby, we named him Michael, is in the premi section of the nursery. We can go there now if you like," Aron suggested.

As they walked down the halls Aron redirected a question to his father-in-law that he had been asked too many times before by his sister, cousins and everyone else that he had called. "Mr. Williams how does it feel to be a grandpa?"

"Exhausting. We rushed to pack, drove 90 miles to Hartsfield, fought through security and barely made it on the plane. We had a terrible flight with people puking all around us, got a Vietnamese Uber driver who scared us nearly to death, could not find the right hospital, was dumped out at an emergency room with gunshot victims, questioned by police, brought to this hospital in a police car, left our bags at the desk and found our way here. No sleep, no food, no nothing. Outside of all that, nothing much happened. That is how I feel about being a grandpa. Let's find a bathroom, and then Aron, let's go see your son."

When they walked to the nursery Aron did not see Michael and in panic repeatedly pressed the call button and a nurse came to the viewing window.

"Speak softly please. The babies are quiet, and I am the only one here."

"Where is my son, Michael Goldfarb?" Aron asked with obvious notes of concern in his voice.

"He was taken out for testing, and was just moved to another spot. I will get him for you."

Returning with a baby wrapped in a blue blanket, the new grandparents saw a wrinkled, pink-skinned baby with almost an inch of brown hair squirming in the nurse's arms.

"He's adorable," Lollie remarked. "He looks just like you Aron.

Those hands – the size of them. He is going to be a big man when he grows up."

"Aron, this is pay-back for all the trouble that you caused your parents," Fred observed. "You are going to find out all about that. What about moving into your new house? Are you all packed?"

"We have cleaned up the new place, but we thought we had another month to finish up. We have gotten some things boxed. Dad said that he was going to call the movers and help us get moved. The Prichards who live downstairs and are buying the apartment also volunteered to help. I am sorry to ask, but could you pack up some things for us? I think that I need to stay here."

Hearing nothing from her husband, Lollie jabbed him in the ribs and Fred reluctantly replied, "I will."

"I'll give you the address. The new house is about ten miles from the old one, but closer to the hospital. We have already had the phone, electricity and water connected."

"My turn now," Fred interjected. "How do you feel about being a father?"

"Scared, frightened, happy and apprehensive all at once. I need to talk to someone to help sort things out."

"There is a bathroom over there," Fred suggested. "Sorry Lollie, guy talk."

Going into the bathroom Aron began, "I can't talk to my dad. He would just tell me to 'Be a man and buck up.' Sam has changed. The last thing she said was 'Get this thing out of me' – like it was an alien. I don't know if she even wants it."

"She hasn't seen it yet?" Fred questioned.

"She was so groggy that I doubt that she remembers. They are going to bring the baby back to see if it will take her breast. She said she wanted to breast feed before the baby came, but I really don't know how that will work out.

"And then there is the sex thing. I have been pulled around like on a yo-yo. She wants it and then she does not want it. I'm afraid that after this she may never want to again."

"Aron," Fred began in as comforting a manner as he could muster.

"You are a worrier who wants everything to go according so some plan. I am sorry, but life is just not like that. It is full of unanticipated events. You will never like it, but you will have to learn to live with it. Everything will work out in the end. You'll see."

Chapter 15

Loading Out

In a large dingy warehouse on the south side of Detroit a black man about 70 years old dressed in a suit and tie pressed the record and answer buttons on his speaker phone as it rang.

"Swift Movers, Alphonso here."

"This is David Goldfarb. My son needs to move into his new house tomorrow. Can you move them?"

"I've got trucks," Alphonso began, "but it's tough to find anyone. They make more money sitting at home. And, if you haven't noticed, the weather's bad and supposed to get worse."

"Whatever your rate is I will pay double. My grandson came a month early, and I need to move them into their new house."

"I usually send four people," Alphonso explained. "Can I tell them that you will tip them an extra hundred? It's going to be tough to get anyone. Not only that, but the Cubs start the World Series tomorrow. I know it's not the Pistons, but love them or hate them everyone is going to want to watch."

"That's extortion," David responded.

"If I am going to get anyone at this time in this weather, I am going to have to offer them something extra," Alphonso rebutted.

"I'll pay a 50-dollar tip," David bargained.

"One hundred or nothing," Alphonso said with a sound of finality in his voice. "I have to look after my people, or I won't be able to get anyone."

"You win," David conceded. "Pick up at the old house tomorrow and deliver to the new house the next day. Here are the addresses…"

The next morning a moving van left the warehouse. On the streets few cars were moving and what few people there were on the sidewalk braced against the wind-driven snow. Inside the cab were Alex Polaskie. Alex, about 45, was dressed in stained blue insulated coveralls and wearing a wool cap with rabbit fur lining and ear flaps. He was accompanied by Henry Smith who was 26 and wearing an oversized set of blue coveralls and wool skull cap under which shoulder-length dreadlocks appeared.

"Once more into the breach dear friend we through peril go to fetch and carry that which we own not to some distant abode through the driving snowy globe," Henry remarked to his somewhat startled companion.

"Droopy Drawers you are on this job because no one else would come. What is this Shakespearean dribble all about?" Alex asked.

"For good sir, I am a Thespian in the Globe Company of Improvisational Theater, and but a bit of character practice I do so not out of easy usage get."

"What's that? A Thespian?" Alex asked. "Is that a lesbian who sucks toes?"

"No," Henry responded. "That would be an interesting roll to play, but I am training to be an actor."

"You don't look much like one, or a college student either," Alex rebutted.

"What say you sir? If I did not wear the garb, I might be set upon by a mob in the place I live where one must ever be on guard.

"How do ye know where we be and go?"

"These trucks have built-in GPS units. I put in addresses, and it provides directions. I need you to look sharp, so I don't run over anyone."

"O flattened pieces of humanity would make we. Red snow cones marking progress to where our destination be. Drive on, and a sharp lookout will I keep."

Back at Aron and Samantha's old apartment David was packing textbooks into an empty liquor box when the door buzzer rang. Looking through the door's peephole he saw a man dressed in camo hunting clothes. After attaching the safety chain and opening the door he asked, "Who is it?"

"I'm Fred, Sam's stepfather. I came to help."

"Sorry," David replied, "I did not recognize you."

David removed the safety chain and Fred stepped into the apartment littered with empty and half-filled boxes in various stages of being packed.

"Aron said that they had started packing, but thought they had another month," Fred stated in a weakened voice.

"How are you feeling?" David asked.

"Terrible. We pushed hard to get to the airport, just made the plane, had a rough ride all the way here, and I am just about to collapse. I need to sleep for a couple of hours."

"The movers are coming sometimes today. Go in the bedroom and lie down. I'll do what I can."

With that remark Fred went into the bedroom, kicked off his shoes and fell on top of the bed. Almost instantly, he was sound asleep. A quarter-hour later the door buzzer sounded again and this time another man appeared that Fred did not know.

"I'm Joe Prichard from downstairs. Aron called and said that he was having to move early and that you might need some help," Joe informed David.

"Yea, and boy thanks." David responded. "I was not looking forward to doing all of this by myself. The movers are coming to load out the house sometimes today. They will take whatever we have ready and deliver it to the new house tomorrow."

"They are not going to like hauling things down those three flights of stairs," Joe commented. "Do you know about the lift beam?"

"The which, what?" David asked.

"There are retractable lift beams over the balconies of these apartments so things can be hoisted up," Joe informed the somewhat surprised David. "They are operated by a crank behind this panel. Mine works, but I do not know about this one. I'll take a look."

Removing a screwdriver from the back pocket of his jeans, he unscrewed a two-by-two-foot wall panel. When he took it off he saw a rat's nest of paper, cloth and leaves which he gingerly took out and dumped into a trash can. Behind the nest was a series of rusted gears on a shaft with the large gear's teeth fitted into sockets on the beam.

"Sorry about that, fellows," Joe said to the unseen former occupants of the rat's nest. "You are going to have to find a new home.

"David. Did you see any oil anywhere?"

"My son is not very mechanical. If there is any, I suppose that it is under the sink," he conjectured.

Joe opened the sink and found gleaming ranks of bottles and cans lined up like tin soldiers standing in ranks. There were powders and solutions for cleaning glass, brass, iron and aluminum; but no oil.

"That's so neat. I hate to pack it up," Joe commented.

"That's Aron. He's always been that way," David replied.

"I'll go down to my apartment, get what I need and come back," Joe said as he opened the door but did not close it completely. As David resumed his packing three mice climbed unseen into a box which David then crammed full of paper-wrapped objects and sealed with several straps of tape.

Joe returned carrying an orange faced mallet and a can of WD-40. He sprayed the lubricant on the metal parts and installed the crank. With both hands he attempted to move the crank, but it would not budge. Taking the mallet, he tapped the metal parts to break the rust bonds and tried again with better results. While leaning on the crank he heard a loud snore and turned to David.

"What was that?"

"That's, ur, Fred. Fred, Sam's stepfather," David informed. "They just got in from Atlanta. He was falling down tired and is trying to get some sleep."

"David could you give me a hand with this crank? I've got it moving, but it needs two people."

With David pulling and Joe pushing the beam began to move with a grating sound which ended with a metallic clank.

"It's hit the cover on the outside of the building," Joe speculated. "It's probably frozen shut. When the movers come, we'll open it."

As David and Joe continue their packing a large moving van pulled up outside.

"Go ring the call button for the Goldfarb family and tell them we are here," Alex ordered.

"Defer, kind sir this request," Henry pleaded. "For me, here, now, I am out of my kin and would likely be shot for thy quest. Call, I pray on thy phone so that a more fitting welcome might there be."

Fumbling through his heavy clothes, Alex retrieved his phone and punched in a number.

Aron, sitting in a chair in Samantha's room, feels a vibration in his pocket and answers his phone.

"University Hospital, Aron Goldfarb."

"This is Swift Movers," Alex replied. "We are outside the building."

"I am still at the hospital. My Dad is in the apartment. I'll call him, and he will let you in."

After waiting a few minutes Alex walked up to the apartment buildings door and pushed the call button for the Goldfarb's apartment. There was a replying "click," and he opened the apartment building's door. He looked up the stairwell and saw the first landing.

"They're on the third floor," Alex observed with a sigh. "Let's go Droopy Drawers. The sooner we can get this done the better."

Both carrying empty moving trucks started climbing the flights of stairs. On reaching the apartment Alex knocked on the door and David opened it. The pair went into the apartment, and David greeted them.

"Glad to see you. The boxes that are ready are over against the wall. We've got a way to get the heavy stuff down."

While Alex is taking off his coveralls, David looks at Henry and stops him from undressing.

"There is a lift beam that we can extend outside of the building, but the cover door is frozen. Can you go outside on the balcony and loosen it?" David asked.

"Tis in sweat I am and pleased would be for a breath of Nature's natural air on my body be," Henry replied.

Joe overhearing this unusual statement from the kitchen where he was packing boxes came into the living room, sees Henry, and asks," Who is this?"

"This is Droopy Drawers. He's a drama student," Alex answered."

"That cover door is outside above the middle of the balcony," Joe informs Henry. "As soon as we can get you outside, you can chill all you like."

Joe, with Henry walking behind, walked across the apartment to the double glass doors fronting the balcony. Joe handed Henry the soft-faced mallet and opened the French doors letting in a blast of snow and cold air.

"The cover plate is in the center of the balcony above the window," Joe stated. "Can you see it?"

"A lift I need," Henry requested. "A stool, a chair, and a strong hand I need to staunch a flying plummet to the street below for I in peril be for feathers I've not even a single one on me."

Alex brought one of the kitchen chairs with chrome-plated legs for Henry to stand on and planted it on the balcony. Then he unreeled some orange strapping tape from one of the hand trucks and strapped it around Henry's chest. Henry climbed onto the chair's seat and using one hand to grasp his safety tape swung the hammer towards the cover plate. A gust of wind caught him in mid swing tipping him over the top of the balcony and leaving him dangling six feet below before Alex's pull and the hand truck's jamming across the balcony's iron railing stopped his descent. With Henry grasping for a hand hold, Joe and David pulled him until he was able to grab the balcony's railing and heave himself over the hand truck and fallen chair onto the balcony's floor.

"Close thing that," Henry commented. "It is a fool who tries the same thing again after so close a peril. Another way I pray."

"Use this strapping tape and tie the handle of the mallet to this broom," Joe suggested. "Then you'll be able to reach it."

On the next attempt Henry stood restrained by two orange straps held by Alex and Joe as he tried to keep his footing on the icy floor of the ledge. Encumbered by his clothing he made four awkward swings at the wall above him releasing snow and ice. Setting the hammer-boomstick combination on the railing the wind whipped it off. It fell like an arrow and centered the windshield of a car parked below which resulted in repeated loud blast of the car alarm. Joe walked over to the balcony's edge, took out his key and shut off the alarm.

"Pitifully sorry I be. It seems that no good deed by me unpunished goes," Henry apologized.

"Couldn't be helped," Joe offered. "Let's get that beam extended before that door freezes-up again."

Joe and David reinstalled the crank and turned it until the beam extended several feet beyond the balcony.

"Now we need a winch, pully and rope," Joe remarked.

"Got those." Alex replied. "There is a winch built into the side of the truck so we can hoist things. I'll shoot a line up to the window."

"Shoot? With what?" David asked.

"A bow and arrow," Alex answered. "Don't worry. Do it all the time. We'll go down, set up the winch and put a blanket over your windshield."

Henry walked towards the front of the apartment and for the first time looked into the bedroom.

"There is a body in here!" he shouted.

"No," David replied. "That's my son's father-in-law Fred. They just flew in from Georgia, and he had to get some sleep."

As Samantha lay in her bed her eyes started to flicker as she struggled to regain consciousness. Aron and the two mothers-in-law walked up to beside her bed, and Aron grasped her hand.

"Aron." Samantha asked. "Are you here?"

"Yes Sam. I'm here. We have a fine baby boy. Our moms are here too. You were so restless in bed that they were scared that you would tear out your IVs and stitches. You have been mostly out of it for two days."

"I think I remember them bringing him in," Samantha postulated. "Are you sure he is all right? I've been having terrible dreams that things were wrong with him."

Lollie came up to the bedside and took Sam's other hand.

"He's fine," Lollie assured. "He weighed over five pounds and came with a head full of hair. All the tests, including the genetics, came back fine."

"I hurt. I hurt bad," Samantha moaned.

"You had a caesarian section, like my cousin Agnes," Sarah remarked. "I'll go get the nurse. She wanted to know when you woke up."

"Tell her to bring the baby after she takes this stuff off me," Samantha requested.

"Honey," Aron began. "I was with you all the way. I got there just before you delivered. Mom was there too, and your mom and stepdad came yesterday."

"Where is he?" Samantha asked.

"He and my dad are at the apartment packing. The movers are there already. They will move everything into our new house tomorrow."

Sarah and the Floor Nurse came into the room. After the nurse recorded the data on the monitors, she took Samantha's temperature and looked at her incision. She asked everyone to leave the room while they washed Samantha and disconnected the monitors. As she worked, she talked to Samantha.

"I'll get you up and walking," she began. "I'll do this in stages. First I'll raise the bed and see how that fields, and then help you to the bathroom."

"I don't know about that," Samantha replied. "I hurt."

"I am going to give you something and come back in a half-hour," the Nurse stated. "Dr. Kitchens will be making his rounds after lunch. He will talk to you about when you and the baby can go home."

Chapter 16

Arrow Flight

Alex pressed a button on the outside of the van's cab and heard a whining noise as a box was dropped from beneath the middle of the van, transported to the outside and lowered to rest on the curb. Alex then opened the wing door back of the seats and withdrew a Kodiak recurve bow and strung it. After he loaded an arrow with a suction-cup point on the front and string on the rear, he called to the balcony above.

"Get away from the window. I'm going to shoot."

Henry bent forward to see what Alex was doing and was hit by the arrow in the middle of the forehead. He staggered back, slipped on the ice and fell onto the living room floor. Joe walked over and pulled the arrow from Henry's head which made a pop as it detached

"Been twice shot and not yet dead, but a bump I'm sure to have on my head," Henry observed.

As Henry rose and regained his composure, Joe threaded the line through the pulley attached to the end of the beam and he and David pulled the line until a cable was drawn through the pulley onto which they attached a hook.

"We'll try it with a couple of boxes first," Joe suggested. "Put them into the net."

THE GOLDFARB CHRONICLES

Henry and David put two boxes into the net and drug it over to the window. Joe and Henry hoisted it over the top of the railing.

"Take up the slack," Joe shouted to Alex below.

With the first two boxes successfully delivered, Alex said that he would come up.

"Empty the fridge, and I'll rig it," Alex said.

Fred, awakened by the noise, walked into the kitchen.

"Sorry David that I couldn't be of any help. These are the movers?" he asked.

"This is Alex and Henry, our movers, and Joe Prichard from downstairs." David said as he made the introductions. "Joe is buying the apartment."

"I'm starved," Fred said. "Can we eat?"

"If we are fast about it," David assented. "Throw some of these frozen dinners into the microwave, and we'll take a break."

The five sat at the dining room table and dug into the TV dinners and washed them down with multi-colored sports drinks. For the first time Henry appeared without his coveralls and his drooping, ripped jeans and heart-printed underwear was exposed. Alex took straps and tied the refrigerator to one of the hand trucks and used ropes to make a suspension harness. The three rolled it to the balcony and lifted it so it was balanced on the railing.

"When I get down and take up the slack, guide it over the rail," Alex said. "When it is free, I will lower it down."

With the refrigerator on the balcony railing; Henry, now redressed, pushed from the top while David and Joe lifted from the bottom so that the truck's wheels could clear the railing. When the wheels rolled off the railing's edge, Henry frantically grasped the ropes as he and the refrigerator fell about six feet and bounced.

"Whoeee. In some peril I be," Henry observed.

"Don't panic," Joe said. "Take a good hold, and let Alex lower you down."

Henry wrapped both arms around the four ropes connected to the cable hook. He then looked down at the swinging earth three floors below, closed his eyes and his lips moved – apparently in prayer as he was lowered through the snow onto the street.

Alex looked up and saw that he had something beside the appliance on this load.

"Droopy Drawers." He greeted. "Nice of you to drop in. Quit horsing around, and help me get that fridge on the trailer."

A little later David and Joe wrapped the sling ropes around the chrome legs of the kitchen table while Henry strapped three boxes of books onto the hand truck prior to taking them downstairs. As the table was hauled to the balcony, Henry opened the door and started down the stairway.

Going one step at the time he started to lose control of his load. He recovered at the first landing and repositioned himself to start going down another flight of stairs leading to the door of Joe's apartment. Henry's foot slipped on melted snow, he lost control of the dolly which rolled down the steps, crashed into Joe's door and burst it open. As Henry picked himself up, he heard the sound of breaking glass inside the apartment accompanied by a blast of cold air and snow.

Henry pulled the hand truck out of the door frame, ran into Joe's apartment, pushed the table out of the French door's broken windows, closed them and pulled up a chair to hold them shut. Henry then rejoined the group in the upstairs apartment.

"Wretched report have I." Henry began. "Slipped on the stairs did I, and loosed like a child running for daylight, the truck with great velocity burst open the apartment door at the instantaneous moment that table legs blown by a fearsome gust dislodged yon doors below, allowing winter's natural gusts to reclaim their natural rule over the abode below with snow, with wind, and with wet."

"I'd better go down and shut things up," Joe remarked.

With no other serious mishaps, the room was finally empty of boxes and furniture. David, Alex, Henry and Joe coordinated the next day's activities.

"You know the address of the new house?" David asked Alex.

"I have it on my GPS," Alex replied. "I don't know when we will get there, but we will start at about eight. I'll give you a call."

"Joe, Alex tells me that you have a four-wheel-drive pick-up?" David asked.

"Yes. I keep it in a garage until I need it." Joe responded.

"We have a piece of furniture that has been in the family for generations that we want to give Alex and Sam," David explained. "Can you pick us up in the morning and take it over?"

"No problem," Joe replied. "I've got tarps and ropes, and there is plenty of room for you both in the truck."

With an ironic round of "see you in the mornings" exchanged between them, the group left the apartment. Henry held one hand on his back and another on his head as he stood on the first landing and looked back up at the apartment door. He shook his head before carefully grasping the rail and easing himself down the steps. He considered not coming back tomorrow, but he would not be paid until the move was completed.

As Dr. Kitchens entered Samantha's hospital room, Aron, Sarah and Lollie were sitting in chairs watching a commentary on the Cub's game. Samantha turned off the TV as Dr. Kitchens walked to the bed.

"It looks like we can send you home tomorrow." Dr. Kitchens began. "The quicker we can get you and the baby out of the hospital the safer you will be from this virus. How did you feel when you walked?"

"Somewhat unsteady, and it hurts," Samantha replied. "I walked down the hall, but I need help getting out of bed."

"The hospital is giving me two weeks paternity leave," Aron informed his wife. "I can help you get around."

"How is the nursing going?" Dr. Kitchens asked.

"A little painful at first," Samantha shrugged. "The baby is hungry when they bring him to me, and he seems to feed well. It's not my breast that hurts. It is the surgical incision and stomach."

"And he pees and shits well too," Aron added.

"We are going to send you home with some formula and diapers. Both are getting hard to find. You will need to pick up some more on your way home. If you want a happy baby keep him fed, dry and clean. You will need to have a car seat before he can leave the hospital."

"I had not thought about that." Aron remarked. "Can I buy one here at the pharmacy?"

"They try to keep them in stock, but I don't know at the moment," Kitchens replied. "Does anyone have any questions?"

"When will we know that the baby is out of danger of that liver disease?" Lollie asked.

"We've run the chromosome test, and there are no genetic defects. We'll take another look at him in a week, and then see him again in six weeks.

"I'll have the baby brought back to spend as much time with you as possible," Dr. Kitchens concluded. "I know that this is an exciting time for you all, but Samantha needs her rest. Tomorrow is going to be a long day."

Later that evening when the room was darkened, Samantha looked at the baby as it nursed and Aron watched, smiling. Aron grasped Sam's hand, bent over them both and gently kissed both mother and child.

The following morning the Speedy Mover's truck pulled out of the garage door and entered a street where blowing snow has now drifted against the curbs and buildings.

Alex had started driving when Henry turned on the GPS unit.

"Not working this is," Henry observes.

"Tower must be down," Alex replies as he hands Henry his phone. "Try Google Maps on my phone."

"Twice foiled we are," Henry observed. "No signal I see."

"Look in the glove box and see if there is a map," Alex instructs.

Henry looks in the glove box and among greasy rags, papers, and tools he pulls out fragments of an old much-stained map.

"Missing parts there are," Henry observes. "Tis far and we must go guided only by some memory star, which I hope burns bright in your head."

"I know roughly where to go," Alex offered, "but there are a bunch of new subdivisions. We will have to look for the address."

Reaching the expressway, they broke out onto a plowed salted road and picked up speed.

"This is better," Alex observed. "Keep trying to call them and get some directions."

Joe's four-wheel-drive backed up into the driveway of Aron's parent's home. After dropping the tailgate, he walked up to the front door and rang the bell.

"Come in. Come in." David welcomed.

"It's getting bad out there," Joe observed. "We need to load up and go."

"Sarah did some cooking and has some things to bring. I'll call my neighbor and his kids to help. It is going to take four people to load this thing.

"Sarah, bring those bags so Joe can put them in the truck."

"This is a Crock Pot of chicken soup," Sarah informs Joe. "I taped the lid on, but it needs to be where it doesn't tip. I have another bag of bread and things."

Now joined by the Goldfarb's neighbor Bill Scott and his two robust teen-aged sons, William Jr. and Bob they go through the open door of the garage and see an antique day bed carved with Egyptian motifs and covered with a worn blue fabric.

"That's neat," sixteen-year-old Bob observes.

"It was built by my great, great, grandfather in France in the 1800s and came to the states with the family," David informed the group. "It has one weak leg that is about to fall off."

"This is beautifully made, although not my first choice in furniture," Joe remarked as he ran his fingers over the carvings of the Gods of Egypt on the headpiece and legs.

Struggling, the four lifted the bed into the back of the truck, covered it with a tarp and tied it down. Although the tailgate was down the bed was restrained by several ropes.

"David, Sarah, are you ready to go?" Joe asked. "I put those bags in the back seat. David, I don't know where the house is. Would you sit up front?"

Joe pulled out of the driveway and the big truck aided by the added weight of its load broke through ripples of snow starting to accumulate on the road.

"Do you have the key?" Sarah asked her husband.

David hunted through his pockets but did not find it. As Joe drove he got out his billfold and removed a shiny key.

"I've got it," David replied with a note of satisfaction in his voice.

"When do you think Aron and Sam will get there?" Sarah asked.

"You never know about those hospitals," David replied. I suspect that we will get there first.

Fred and Lollie finished their breakfast at the bed and breakfast which was located a few blocks from Aron and Sam's new apartment.

"It's nine," Lollie observed. "Do you suppose that anyone is there?"

"I don't imagine so," Fred postulated. "We have a key so we can walk over anytime. It's just a couple of blocks."

"I don't know if we brought enough clothes?" Lollie questioned. "The weather is frightful."

Their host Karla Karoboskie walked in with a steaming pot of hot coffee.

"Mam," Fred said. "We have to walk a couple of blocks and didn't bring heavy enough clothes. Do you have some coats that we could borrow?

"I do," Mrs. Karoboskie offered. "They're old, but they are warm. They are Red Army wool coats that my father and uncle wore during The Great Patriotic War. They have seen snow and cold before."

Two bent-over figures in brown coats scurried over the sidewalk with snow sticking to their coats, scarves and hats as the larger of the two reached into his pocket for a key.

"The key is in here somewhere," Fred stated.

He reached into the cavernous coat pocket, got out a package of matches and returned it. Then came a crushed cigarette pack followed by a ticket stub. He pulled the inside of the pocket out and the key fell with a metallic tink as it hit the concrete. He reached to get it but fell and caught himself on his hands and knees. Lollie helped her husband up, bent over and retrieved the key.

"This won't work," She observed. "This door has a key pad. Do you remember the number?"

"It's 6743," Fred replied.

Lollie punched in the number and opened the door. Inside was a stairway leading up to the apartments on the second and third floors. A door to the left opened into the garage which also contained with laundry room and storage cages for snow tires and winter car accessories. Fred looked at a screen on the garage door's control panel and saw a pick-up backing up the drive towards the door. Fred spotted the door open button and presses it. The door opened and Joe backed his truck into the garage.

"I'm glad to see you," Fred said. "Lollie and I just got here"

"I need to get unloaded and leave," Joe remarked hurriedly. "That snow is starting to clog up the roads. I have brought David and Sarah with me."

As quickly as they could they took the tarp off the truck and as gently as possible all assisted in lowering the day bed to the floor.

"I don't know why, but this thing gets heavier every time I move it," David said.

"We'll take the small stuff up to the apartment, and let Joe get out of here," Fred stated. "That snow is not getting any better. Joe, I really want to thank you for all of your help. I'm sorry about your windows and stuff."

"Me too," Joe replied. "But when you have 'adventures in moving' these sorts of things happen.

Chapter 17

Unexpected Passenger

The moving van slowly made its way through a series of secondary streets and stopped at a cross-street. Henry got out to examine a street sign. As he brushed off the sign, he heard a whimper coming from beneath a cedar tree. When he parted the tree's limbs, he found a dog tightly curled up beneath it. He reached down and petted it.

"What's taking so long?" Alex shouted against the wind. "Is that Springdale Circle?"

Henry bent over, picked up the dog and cradled it in his arms as he brought it to the truck.

"A creature in desperate need of aid have I found under yon sheltering boughs," Henry pleaded. "Take it with us we should, for I fear twill not last the night."

"We don't have time for this," Alex replied.

Henry spotted a pick-up truck making its way down the street and shouted to Alex, "Look. A truck."

As Alex was distracted by the oncoming four-wheel-drive, Henry put the dog on the seat, got inside and closed the door. The pick-up came abreast of the driver's door, and Joe rolled down his window to speak to Alex.

"Follow my tracks," Joe suggested. "I just left the apartment. It's about a mile away."

"Right on," Alex replied. "My GPS won't work. Phones neither."

The dog reached over and licked Alex's hand. Alex reciprocated and scratched the dog behind the ears.

"To save a life, can she ride?" Henry asked. "Look after her I will, and food and lodging provide."

"I am an old softy, I guess," Alex responded. "But Droopy Drawers you damn well better look after her or out she goes, blizzard or no blizzard."

Following Joe's truck tracks through the snow-bound streets they found the apartment and Alex backed the van into the driveway in front of the garage door. He got out and rang the button for the third-floor apartment.

"Hello," Fred responded.

"Can someone open the garage door so we can back in?" Alex asked.

"We will be right down," Fred eagerly assured the driver.

Fred punched the keypad and the garage door opened allowing the van to partly back into the garage. Alex got out and approached the group which now also included David, Sarah and Lollie.

"As soon as we can get unloaded, we can pull out and close the door," Alex observed.

After greetings were exchanged everyone squeezed into the elevator. David pushed the button for the third floor and the elevator rapidly ascended and abruptly stopped.

"Is everyone all right?" Lollie asked. "That took the breath out of me."

The elevator door opened, and the bottom was six inches above the floor, so everyone took a half-step down to the hall floor. Lollie opened the apartment door. The group found a combination kitchen-dining-living room, two baths, a master bedroom, two smaller bedrooms, pantry and closets. One of the smaller bedrooms had a fold-down Murphy Bed built into the wall.

"I'll find some lights and turn up the thermostat so we can have some heat," Fred offered.

Inside the front room he discovered a wall panel with a variety of gages and buttons. He turned one red nob and heard the satisfying sound of air rushing through the heating ducts. The family members removed their heavy clothes and hung them in a closet.

"If we know what's in the boxes we can sort out where they need to go," David remarked. "The corner bedroom with the built-in desks and Murphy Bed is the office. The bedroom with the adjoining door to the Master Bedroom is the nursery and the kitchen area is against the back wall."

"There are a lot of shelves for putting out their keepsakes and dishes and even a pantry," Lollie observed. "This is a beautiful airy room."

"Alex, if you can bring in the big kitchen stuff first, I'll start hooking it up," Fred volunteered.

"What do we do with that old bed?" Alex asked. "I need to get it out of the way."

"Put it in front of the window overlooking the street for now," David suggested.

"Good folks all, a necessary body function I need to discharge," Henry stated with a hint of embarrassment in his voice. "Where go I?"

Lollie dug into her handbag and pulled out a packet of tissue papers. "Take this young man until we can find some toilet paper. There is a bathroom down from the fireplace."

Alex, David and Fred with the aid of the hand truck moved the day bed under the window. The three took the stairs to the first floor landing and then into the garage where they proceeded to unload the van. Once emptied, the van was pulled out on the street and the door closed. While they were attaching the hand truck to the refrigerator the door opened and Aron and Samantha drove in. David and Fred approached the car.

"What took you so long?" David reprimanded Aron.

"It took forever to get released from the hospital," Aron began. "Then there was a wreck on the road. Then we had to pick up some diapers and baby stuff and most of the pharmacies were closed."

Aron got out of the car and David ordered, "Close that door before we freeze to death."

Aron fumbled with the fob on his keychain and ultimately located the correct button and closed the garage door which shut with a crunch against the ice and snow.

"Sam," Fred said with concern. "Let's get you and the baby upstairs. We were starting to worry about you."

Fred helped Samantha out of the car and walked her over to the elevator and punched in the code. The door opened and they were joined by David and Alex. As they stepped inside the elevator door closed with a grating sound.

"Hang on Sam, this elevator seems to have a mind of its own," David advised. David pushed the button for the second floor, and the elevator began a measured assent to discharge its passengers perfectly at the second floor.

As Samantha entered the room, Lollie observed. "Sam you must be exhausted. There is a bed in the corner bedroom. Haven't found sheets yet, but we have some coats that you can rest on."

Lollie and Sarah pulled down the Murphy Bed and brought in coats from the closet. Sam is put to bed with a folded Red Army overcoat as a pillow and another as a blanket. Nestled and pillowed in the arms of the Red Army, Sam and the baby fell asleep.

Alex and Henry pushed the hand truck with the refrigerator to the kitchen wall where Fred and David waited.

"There is a plug and drain connection here, but I don't have any tools," Fred remarked.

"I have a toolbox in my truck," Alex advised. "I'll bring it up with the stove. You know you need a Union plumber or electrician to do that?"

"If we waited for union labor, we might be a month," Fred augured. "Just bring the tools, okay?"

A few minutes later Henry and Alex returned with the electric stove and a rusty tool box that Alex bangs down on the floor. "These are my tools," Alex said. "You make sure I get them all back."

As Alex and Henry unstrapped the stove, Fred examined the appliance's plugs.

"These 220 plugs are different for the stove," Fred observed. "I am going to need to splice the wires together to get them to work."

"Do what you want," Alex told Fred in something of a huff. "If you burn this building down because of what you do, that's no skin off my nose. It's you that is going to jail."

"I think that he's right," David agreed. "If we need to heat something we've got the microwave. I'll get the electrician back to wire up all the appliances."

Lollie, sensing that things were starting to get heated between the two men interrupted. "Aron, is the baby stuff still in your car? I'll help you get it organized. He is going to need changing soon."

As Alex and Henry took the hand trucks down for another load, Aron and Lollie talked.

"They gave us a few infant diapers at the hospital, but all the pharmacies were closed. We found a Chinese all-night place and they sold us a box of used cloth diapers, some wipes, condensed milk, Karo syrup, a half-dozen glass bottles and a box of nipples. All of these need to be washed and sterilized."

"Cloth diapers?" Sarah questioned. "I didn't even use them on you. Did you get pins?"

"I've got some safety pins in my bag," Lollie offered. "Let's go down to the utility room and wash those diapers. I know someone who is going to need them soon."

In the garage Lollie, Sarah and Aron unloaded boxes and bags from the car and took them to the laundry room or elevator. Lollie noticed a packing blanket on the floor with an indistinct form lying on top of it. When she approached for a closer look, she exclaimed in surprise, "It's a dog!"

Aron and Sarah went over, and when they approached the dog wagged its tail thumping the blanket. By the time the others arrived, Lollie was stroking the dog's fur.

"Where did that come from?" Aron questioned.

"It's on a moving blanket and someone put down water for it, so I suppose the movers brought it," Lollie conjectured.

The door to the elevator opened and Alex and Henry came out with empty hand trucks for another load. Henry walked over to the group.

"Yon bitch, that canine, is mine discovered on our journey

abandoned, alone and cold. Taking pity, I took it and deposited it here for better protection from the cold than in yon metallic steed."

"This dog looks terrible," Sarah observed. "Did you know that it is about to have pups? Did you give it something to eat?"

"I gave it a can of Vienna Sausages," Alex responded.

"Lift it up and bring it over to the sink." Lollie requested. "I want to wash it if it will let me. And find something more for it to eat."

"Unless you want this dog having her pups in your truck, it needs to stay here," Sarah observed. "I'll take care of the dog and get it back to you if you want. This dog and the pups can't be out in this weather."

"Kind lady," Henry responded. "One dog bargained I. Not three or four or five for meager resources for food and gentle care must be applied. I accept your offer and blessings be applied."

While Alex and Henry continued moving the household materials, Lollie, Sarah and Alex start washing and drying the diapers and washing Momma Dog who is dried with towels and a hair drier.

Open boxes, newspaper packing, cardboard and household items littered the floor as the last of the cartons were delivered upstairs.

"That finishes us up," Alex said. "Someone needs to sign these documents because we need to get out of here."

"Aron should sign," David responded. "This is his house. Where is he?"

"Down with the women washing diapers," Alex responded.

"Well, maybe that is the best place for him," David answered with a note of sarcasm in his voice. "I'll sign."

David takes the clipboard containing the documents and signs it returns it to Alex. Alex stands and then holds his hand out.

"O yes. Your tip."

David reaches into his billfold and pulls out two one-hundred dollar bills and puts them into Alex's hand.

"That should be $400," Alex demanded.

"I am paying double the usual rate and one-hundred dollars a man extra," David retorted.

"The usual crew is four and Henry and I did the same amount of work. Out tip should be two-hundred-dollars each not one-hundred dollars."

David reached into Alex's hand and took back the two-hundred dollars.

"Two hundred or nothing," David replied.

"You are a skinflint," Alex commented through his teeth. "I am sure your generosity will be rewarded."

A scream followed by the sounds of a baby crying are heard from the quest bedroom. The men rushed inside and found that the Murphy Bed had almost folded itself back into the wall with Samantha and the baby trapped inside. Fred and David pulled the bed down and helped Samantha sit upright so that she could hold the crying baby in her arms.

"The baby nursed, and we went to sleep," Samantha explained. "I felt as if I was being smothered. I was trapped and screamed. I have never been so scared."

"I guess that bed is regulated for one or two normal-weight people and folded up because you were lighter," Fred postulated. "It didn't know you were there."

"Well, I sure as hell knew I was there," Sam retorted. "Where are those diapers we brought from the hospital?"

"Aron, Sarah, and your mother went down for them an hour ago," Fred explained. "I don't know what's keeping them. I'll go down and see."

About fifteen-minutes later the apartment door opened and Fred walked in with a box of diapers. Sarah and Lollie followed with each holding paper bags full of baby items. Aron with the last to enter with Momma Dog wrapped in a towel.

"Get that filthy animal out of my house," David ordered.

"Dad. Calm down," Aron interjected. "First this dog has been bathed and is very clean. And secondly, this is not your house. This is Sam's and my house.".

"The black guy found it in the snow and brought her here," Fred explained. "It is about to have pups and none of us could put it out or leave it downstairs."

"It is going to have its pups here," Sarah explained. "And we, husband dear, are going to take it home until the pups are old enough to give away. Meet your new family."

Aron carried the dog to a closet and laid the dog on its shipping blanket on the closet floor. He then went to the sink, filled a bowl with water and set it on the floor beside the dog.

"How are you and Michael?" Aron asked.

"We nearly got eaten by the Murphy Bed," Samantha answered. "It folded up on us and your and my dad had to get me out. And your son needs changing."

"I can do that," Aron assented. "I got all the stuff."

On the kitchen table Aron laid out a tower, rubber gloves, baby wipes, lotion, a cloth diaper and a pair of safety pins. After washing his hands and putting on his gloves, he took him over to the table where he removed the diaper, sealed it in a plastic bag and threw it into a plastic tub which he pressed down on until it locked. Returning to the baby he wipes his bottom, rubs it with lotion and places him on the tri-folded diaper. Now ready to pin him up the baby's penis stiffens and spouts like a geyser wetting the diaper beneath. Nonplused, he dries off the baby, applies more lotion and placed the towel and wet diaper into the laundry tub and sealed it.

"See. I got this," Aron proclaimed with a smile on his face.

Chapter 18

Marooned

Once the documents were signed, Alex and Henry began their trip back to the city. After bucking through ripples of snow a foot deep, three blocks down from the apartment they were confronted by a drift that was five-feet high.

"The smaller stuff I can bust through, but I can't make it through that much snow," Alex observed. "It is supposed to 20-below zero tonight, and we can't stay in the truck. We are going to have to go back. Get out and help me turn around."

With Henry giving directions, Alex avoided hitting any cars, but did take out a mailbox. After moving up and backing a few inches at the time, Alex managed to get the truck turned around.

A series of knocks was heard on the door of Aron's and Samantha's apartment. Aron got off the couch to answer the door. When he opened it he is startled to see Alex and Henry.

"The street's blocked with drifts," Alex informed, "and we can't get out."

"We are snowed in?" Aron asked disbelievingly.

"By icy drifts a half-fathom high and 50 degrees of frost in winter's ice grip are we marooned like sailors caught in the Arctic's icy grip," Henry remarked.

"It is too dangerous to go outside in this temperature," Alex explained. "We can't get out until they plow the road – maybe day after tomorrow."

"There is not enough room," David opined.

"There's room for everybody," Sarah rebutted. "We just need to get organized. There are three beds and the couch, counting the convertible crib in the box. Fred and Lollie should stay too."

"We have room enough," Aron agreed. "There are also enough blankets and sheets. It's food that we are short of. Sam and I ate much of what we had in the apartment so we did not have to move it."

"I keep some canned stuff in the truck," Alex volunteered. "We fed some to the dog, and I don't know how much is left."

"I have that soup that I made and the bread I brought," Lollie added. "I also have some crackers and maybe a candy bar from the airline in my bag."

"We can always order in from Pizza Hut," Aron quipped which was followed by general laughter. Not even Pizza Hut was going to deliver in this weather.

"If we move these boxes and break down those that we have emptied we'll have more room," Fred suggested. "Aron, Sam, tell us what goes where?"

"Since we did not pack them all," Aron replied. "I don't know until we open them."

"I know what some of them are," Samantha said. "Those four flat boxes go into our bedroom. That square one, The Snoo Bassinet, can stay in here for now."

"Are you two guys just going to stand there?" David demanded. "You're movers aren't you?"

"You are moved," Alex replied forcefully. "We have discharged our contract and you have signed off on it. For anything else you need to pay us."

"You didn't give them their tip?" Sarah asked. "Did you?"

"They wanted four-hundred dollars," David sheepishly replied. "I gave them two hundred."

Sarah reached inside her handbag and pulled out an old-fashioned coin purse and from it extracted two hundred-dollar bills and handed one to Alex and another to Henry.

"Will this help?" Sarah asked. "You children play nice. We are going to be stuck here for who knows how long."

"Thank you," Alex replied and then turned to Henry, "Clear a path to the bedroom and take those boxes over there."

Henry picked up two of the smaller Ikea boxes and walked toward the master bedroom moving boxes out of the way with his feet and body as he went. Fred and David moved others to near the pantry and kitchen. Sarah unpacked the kitchen things while Lollie put items on the shelves. As the men brought her boxes and opened them, Samantha sat on the sofa with the baby directing operations.

"Unpacked I have device wondrous strange and here its electronic brain I bring," Henry remarked. "To what aim and how works, doth it?"

"That is the Snoo Smart Bassinet," Samantha explained. "The baby sleeps in it and when it cries it rocks. It also holds it on its back so that it can safely sleep. If it cries too long or moves too much, it will ring the control so that someone can see about it. Put it here and I'll put the baby to bed."

Henry put the cardboard against the growing pile of packing materials against the fireplace and retrieved the bassinet. Everyone gathered while Samantha put the baby inside, wrapped the swaddling wings around its chest, zipped it inside its cocoon and turned it on. The bed moved the baby gently from side to side while emitting a soothing sound.

"Sleep to that I could," Henry remarked.

"I wish I had that when you were born, Lollie remarked to Samantha. "You didn't want to do anything without squalling about it."

"Aron, your dad and I lost a lot of sleep when you were this age," Sarah informed. "How long can he stay in it?"

"Dr. Henry Karp who wrote this book on infant sleep, says up to six weeks," Samantha replied. "Everybody I've talked to said that it works, but as big as Michael is, he is going to outgrow it in a hurry. I don't think that we are going to have that problem, but some have reported that it was difficult to get them to sleep in a normal crib after they outgrew the Snoo."

"Speaking of growing," David said. "That crib that we bought you can be expanded into a toddler bed, an adult bed and ultimately a sofa. Sarah and I thought it would be just the thing for you. We can put it together. Where do you want it?"

"In the nursery so we don't have to move it again," Samantha responded.

"I'd like to see how it works," Fred responded. "Alex, if I may borrow some tools, I'll help."

"I'll work on the chest of drawers in the bedroom," Lollie stated.

The designated furniture-assembly crews went to their respective areas. Alex and Henry with box cutters helped open the boxes and laid out the parts as they went from room to room. With the parts exposed on the floor, it looked like a factory assembly line. Lollie was laying out the parts of the chest-of-drawers and Henry came in.

"Come to help I have," Henry offered. "Methinks to align pins and holes more hands that a pair needed be."

Lollie, on her knees, worked on the frame while Henry opened and sorted the parts for the six drawers. As he moved his pants dropped lower and lower across his butt and fell around his ankles exposing his underwear and bare legs. Embarrassed, he dropped his tools and pulled up his pants. Lollie, hearing the noise, turned around and observed his efforts. She reached into her bag and pulled out a coil of Mule Tape and handed it to Henry.

"Young man, you need this," Lollie observed.

"For modesty, decorum and practicality I do," Henry admitted. "Thank you again for your kind attention to my plight."

Henry threaded the tape through his belt loops, tied a knot in it as Lollie retrieved blunt-nosed children's scissors from her bag. Henry nodded and reached into his pocket and withdrew a box cutter, cut the Mule Tape and returned the remainder of the roll to Lollie.

"Now, with trousers thus restrained no longer Droopy Drawers, am

I," Henry asserted. "Now I can be revert to my rightful name, Henry, and by thus be further identified."

"Glad to meet you Henry," Lollie responded. "Hold these pieces while I screw them in."

Alex brought boxes for Aron to open in front of Samantha. As Aron opened a box a trio of mice escaped and ran through the boxes towards the closet and shelf-lined wall adjacent to the door. Aron jumped back in surprise. Momma Dog saw the running mice and gave chase scattering empty boxes and loose papers across the floor.

Samantha shrieked and cried, "A rat!"

Everyone in the apartment came into the room and attempted to herd the mice towards the door. As the chase circulated through the room Aron opened the door while securing it with the chain making a potential escape route. The mice escaped through the crack, and Momma Dog stood at the door panting. After receiving pets from Aron, she held her tail high and trotted back to her bed in the closet.

"That was exciting," Sarah said. "Let's have some food. Put what we have on the table, and I'll put out some dishes."

Sarah set the table for eight. The soup and bread had already been consumed for lunch. Lacking that, she sat out seven scrap plates, forks and napkins and a dinner plate for the last TV dinner, artichokes and anchovies with rice, for Samantha. The remaining plates had two Vienna sausages, two saltine crackers, a slice of a Snicker's bar, tablespoon of green beans and for Momma Dog, a half of a can of dog food recovered from the back of the pantry which Henry took to her.

"Fair hound," Henry addressed. "You dine better that we this day. On my return out into the blustery weather we go for necessities you do."

As Henry returned to the dinner table Fred had a conversation in progress "…the Donner party and those trapped miners would up eating their dead. I remember hearing that in Jamestown…"

"We are only going to be here for a few days," Lollie reminded everyone. "Outside of Henry here, I don't think a little fasting would hurt any of us. Henry, you are not going to wind up on the menu."

"Most grateful I am, for tough, stringy and long cooking I'd be.

Tis more harm than good I would be. I will dress, and dog and I will venture forth, for fed she's been and eliminate she must."

Turning to Lollie Henry asks, "May I from you have more cord for a collar and leash to fashion?"

Lollie reached into her bag and produced the roll of Mule Tape from which Henry makes his dog leash.

Henry, dressed in his coveralls, with Momma Dog on her leash, went down in the elevator which progressed in a shuddering manner. He reached to pet the dog and heard voices speaking a language that he did not know.

"Thy spirits, if that is what you be," Henry entreated. "Speak in a tongue I know or silent be."

Henry's ride to the garage continued in silence. He then opened the garage door letting in the snow and a raging wind. Pulling down his cap tightly on his head, he and Momma Dog walked outside.

"Be brave, good dog," Henry encouraged. "Do what must be done."

Momma Dog hesitated and then ventured into the drifting snow.

Following three raps on the door, Sarah with a towel in hand went to the door and let Henry and Momma Dog in. She used the towel to wipe the snow off the dog's back and then Momma Dog rolled over to allow the melting snow to be dried from her underside.

"Good dog," Sarah responded.

"Twas quick. Twas Fast. Twas cold," Henry reported. "Notwithstanding wherewith all, she did what was needed."

"What is the snow doing?" Alex asked.

"Frozen ice cream it be on the ground and now unrestrained powder drifting across a three-dimensioned canvas. Tis deep and deepening, it is."

"There is one thing more."

"What? Alex asked.

"Spirit voices in yon shaft hear I," Henry replied hesitantly. "Not with threat, bluster, or malice; but talk as you and I."

"Kid," Alex replied. "Whatever you are smoking or taking you had better lay off. That stuff is rotting your brain."

Henry pulled the improvised collar from the dog's head and Momma

Dog returned to her bed in the closet and lay down as Henry pulled off his coveralls.

"How are we going to sleep?" Fred asked.
"I sleep naked," Alex replied.
"I mean there are six of us and we have three beds and a sofa."
"On a floor, on a wall, in the hall will I sleep," Henry remarked. "But not in a bed with a naked white guy, no offence intended or ill practice implied."
"None taken," Alex responded.
"Does that old bed work?" Fred asked.
"We've used it a few times since we were married," David informed. "And Aron, you were conceived on it. That's all we had when we married."
"To much information, Dad," Aron interjected.
"I noticed that it had a loose leg when we were moving it," Fred said. "It's missing a cross-piece. If someone has a ball-point pin I can shave it down to fit."

Sarah retrieved her handbag, rummaged in it for a few seconds and produced a ball-point pin from the Circus-Circus Casino in Las Vegas.

"You remember David," Sarah asked. "We went there on our tenth wedding anniversary. They still had coin-operated slot machines."
"I could not play them," Aron said. "I remember that Dad lost his roll of quarters and would not touch them again."
"Damn right," David replied. "If I'm not winning, I'm not playing. Just good business."

"Ur. Henry, is it?" Fred asked. "Come with me and I'll see if we can't fix that sofa."

Walking over to the sofa Henry discussed the Egyptian figures. "There I Ammon-Ra in the east, the protective cobra and vulture in the middle and Anubis on the end. The legs walk with toothy Ammut on one side and the lion-headed Isis-Mehtet on the other. Well protected will I be by the ancient gods of Egypt."

"How do you know about them?" Fred asked.

"It was I, kind sir, who helped build the Anthony and Cleopatra set, and did act in that play doing quintuplet parts."

Fred used a pocketknife to shave some plastic from the barrel of the fountain pin, screwed on of the rear legs back into place and tapped the pin into place leaving a part of the pin sticking out so that it could be replaced by a better fitted part.

"I can see why the Goldfarbs hung onto this," Fred Admired. "It might look ugly as sin, but it is a magnificent piece of craftsmanship. There are few indeed who could equal this type of work today. Help me pull it out.

"Except for the tacks holding the cloth onto the frame, there is not a piece of metal in the entire thing."

"Lumpy and bumpy it be," Henry observed.

"Sweep it off and fold one of your moving blankets over it, and it will be fine," Fred advised. I know there are boxes of sheets and blankets somewhere although we may be short of pillows."

Chapter 19

Neighbors

After the conclusion of a very long day, Samantha and Aron made it into bed with Michael lying quietly in the bassinet beside them.

"Sam, we and Michael are finally in our new home. I know you are exhausted, and I am too. We need to talk."

"About what?" Samantha asked.

"I think you know," Aron replied.

"Know what? Samantha queried.

"You know," Aron responded.

"Oh. Him," Samantha agreed.

"You use to call him Sir Lancelot and carry on about him. But he has not had any attention for more than a month."

"You know we can't with this genetic stuff," Samantha rebutted.

"With a raincoat on we could, like we did before when you wanted him so badly," Aron pleaded.

"Tell him to go back to sleep," Samantha instructed.

"He is not going to like that," Aron argued. "He may stop working at all."

"That would be fine with me," Samantha concluded.

Michael started crying. He had apparently been disturbed by the conversation.

"See," Samantha said. "You woke the baby. Go see about him."

Aron got of bed, took Michael from the crib and went into the bathroom to change him. After he returned, Aron sat on the side of the bed, rocking his son in his arms. He kissed him on his head, put him back into the bassinet and got back into bed beside his sleeping wife.

Henry slept on the daybed with a sheet and blanket pulled tightly around his neck. His limbs moved under the covers as if he were running. In his dream he was wearing droopy trousers and a T-shirt while running up the steep face of a sand dune. He fell to his knees and looked back to see an army of black jackal-faced warriors from the movie "The Mummy" pursuing him. Overhead a vulture swooped down with its claws outstretched and an open beak to peck out his eyes. Henry threw up his arms and made a guttural sound.

Momma Dog heard the noise, got up from her bed and walked over to Henry. She put her paws on the side of the daybed and licked Henry in the face. Henry woke up, startled. He then realized where he was and petted the dog.

"Good dog," he commended. "Thank you. You knew that was bad, didn't you?"

Satisfied that she had done her duty, Momma Dog responded with a wag of her tail and returned to her bed.

Snow pelted the outside window, but the living room was somewhat illuminated by light coming through the windows. Henry was still sleeping when Alex dressed in his jeans, shirt and shoes walked over to him.

"Rise and shine, Droopy Draws," Alex remarked. "It's time to start the day."

"There are days to do and days to not do. This good sir is a day to not do," Henry argued. "Pelting snow has beat a steady drum all night on yon window. If slackened, I know not. If worsened, I could not tell. But snow, snow and more snow has fell."

"Snow or no snow, you have a dog to take out," Alex ordered.

"A debt I owe to dog," Henry explained. "For she from my somnolent terrors rescued me with a favored lick from creatures that I could not describe, save of ancient Egypt they were derived to stalk the valleys and ridges of my furrowed brain."

Fred came out of the bedroom that he and Lollie shared and walked over to the pair.

"Do you have anything else to eat in the truck?" he asked.

"No." Alex replied. "I brought it all in yesterday."

"I guess that it's going to be Gator Aid for breakfast. There is nothing else in the fridge," Fred reported.

After getting up and putting the improvised leash on Momma Dog, Henry took the dog down the elevator to the garage where he opened the door. He had to shovel drifted snow to make a path for him and the dog to walk out into slightly more open areas in front of the house.

"There you are. Go on," Henry instructed.

Hesitating, Momma Dog tested the snow with a paw and then walked out to a less drifted-in area to discharge her biological functions. When she returned, Henry and Momma Dog waited for the elevator to return and the door to open. When it did, there was a cardboard box piled high with cans and sacks of food on the elevator floor. Written in bold letters with a Magic Marker was "Welcome to Springdale."

"What this I see?" Henry said in surprise. "Food in plenty for hungry mouths and slack bellies there be."

When he and Momma Dog arrived at the apartment the box was put the kitchen table and its contents examined.

"We have this big canned ham which is cooked, two cans of Spam, flake instant potatoes, dried beans, rice, canned corned beef, beets, guavas, string beans, corn, butter beans, flour, canola oil, a loaf of bread, noodles, chicken stock, coffee and a pack of cigarettes and matches, although I don't think any of us smoke," Sarah enumerated.

"There is a lot that I can make from this with the microwave," Lollie observed. "How about some mashed potatoes and corned beef for breakfast?"

All agreed and Lollie began to assemble their meal.

THE GOLDFARB CHRONICLES

"Where is that coffee maker?" Fred asked. "I'll make some coffee."

"We have a toaster. I can make some toast," Aron volunteered. "Sam, what do you feel like for breakfast?"

"At this stage I don't care," Sam responded. "I feel like I could eat anything."

"Where did this stuff come from?" David asked. "Henry did you see anyone or any tracks outside?"

"Most curious this is," Henry replied. "Outside just fresh, drifted snow. No transport came and no one passed on the general way nor had there been all day."

"Aron, who else is in the building?" David asked.

"No one that I know of," Aron puzzled. "The older couple who lives below us went to Florida weeks ago."

"I have seen some curious things in my life, but not a box of food appearing from nowhere," David observed. "Someone must have put it in the elevator."

"Accept what you have and be grateful for it, Sarah admonished. "All will be revealed in God's good time."

Everyone is sitting at the table when a flushing sound is heard.

"Someone else is in this building," David observed. "Where did that noise come from? It wasn't any of us."

"Aron, let's you and me check the apartment downstairs," Fred suggested. "If that couple is gone, someone may have broken in. Do you have a gun?"

"No. I have never owned or shot one," Aron replied. "They are obviously trying to help us, why would we need a gun?"

"Guns are something that I need to teach you about," Fred replied. "In my day every guy who could walk was going to do some service time. Let's go."

After having gone down to the second-floor apartment Fred repeatedly rapped on the door but received no response. Looking at the landing in front of the door they saw that the only footprints on the dust were their own.

"It does not look like anyone has been here for some time," Aron observed. "Certainly not since it started snowing."

"We need to go outside to check the windows," Fred suggested. "Someone would have to have a ladder to reach them."

After their examination of the outside of the building they returned to the apartment with their snow-dusted clothes. They took off their shoes on the flagstones at the entrance. As they undressed, Fred gave his report.

"We looked inside and out. No one has been in that apartment. There are no lights on and no signs that anyone has gotten in through a door or window."

Pointing upwards, David asked, "How do we get up there?"

"There is a key for an elevator slot that allows access to the attic, but I don't have one," Aron responded.

Lollie reached into her handbag and pulls out a large ring of keys.

"Does it look like any of these?" Lollie asked.

"It would be a round one," Fred informed. "Like a refrigerator key with nobs on the inside."

"I've got two of those," Lollie informed the group.

Aron, Fred and David went inside the elevator and Aron tried the first key with no success. When he used the second key, the elevator hummed. He then pushed the up button and the elevator started a smooth assent to the attic. It stopped and the door opened.

The attic was a sparsely furnished unfinished apartment with pressboard walls and floors, electricity, plumbing and appliances. Light was provided by windows in the gables and four small windows in dormers projecting from the roof. When the door opened, the group saw Erderlan Pasha, his wife Keje, and boys Bryan and Tom. Erderlan walked to the elevator door to greet his visitors.

"Come in and have some tea," Erderlan invited. "I was wondering when you would find us."

"I'm sorry," Aron replied. "I didn't know anyone else was in the building. I am Aron Goldfarb. This is my father, David, and father-in-law Fred Williams."

"Keje, Brian, Tom come and meet our new neighbors," Erderlan instructed.

As they approached Erderlan introduced each of them in turn.

The boys approached shyly. As they had apparently been taught, they reached out their hands to shake the men's hands.

"Glad to meet you ... sirs?" Ten-year old Bryan spoke in understandable, but somewhat hesitant English.

With a nudge, eight-year-old Tom also approached the men and extended his hand without speaking.

"The boys are having to get use to speaking English," Keje explained. "We are trying to home school them here with computers, but they don't get to talk to hardly anyone else but us."

"What are you doing here?" David asked.

Erderlan looked at David and ignored the abruptness and potential hostility of the question.

"I was a third-year medical student when war broke out in Syria," Erderlan began. "We are Kurds, and I helped in hospitals during the fighting. Then we fled to Turkey, Cyprus, England and finally to the United States. A refuge organization found this place for us. Keje and I are working on our citizenship. I'm an Uber driver until I can complete my medical training, and Keje works as a translator and English teacher."

"With all those moves, I don't wonder that the boys are confused about languages," Fred observed.

"It was you who put the food in the elevator?" Aron asked.

"We could hear what was going on downstairs and felt like you could use some help," Erderlan stated. "We gave you some things that we are forbidden to eat and some others that we did not know how to cook."

"You are Muslims, and you shared your food with us?" David questioned.

"We are all people of The Book," Erderlan explained. "It is my obligation to help others in need when I can."

"When it is convenient," Aron suggested, "come down and meet everyone."

"Perhaps later in the afternoon when Bryan and Tom are finished with their lessons," Erderlan responded. "The internet is down, but we have some programs on our laptops that they are working on."

When the men returned, Lollie and Sarah were interested to learn about the new neighbors.

Fred began, "They are a young couple. Kurds I think he said with two boys. He was studying medicine in Syria, and they were caught up in the war. They moved all around the Middle East, and somehow made it here."

"His wife is working as a translator and he is picking up some money as an Uber driver," Aron continued. "He told me their names. They were something like Ur-de-lan and Key-gee. The boys are Bryan and Tom. They speak excellent English, and the boys are learning."

"They are Moslems," David added. "Those are the first Moslems that I have ever really met. They shared what they could not eat, but we can't either."

"There are other things in that box besides pork," Sarah reprimanded. "They gave freely of what they had without expectation of reward - like we are supposed to do."

"Methinks that friendly hand they've extended and friendly grasp we should return," Henry stated. "For staunched our protesting stomachs they have with what little they had. Twice blessed are those who give and those who receive in a kindly spirit."

Those appropriate words from Henry elicited looks of surprise from everyone.

"I did not know you were a philosopher too," Aron responded.

"The Bard in his plays has many life lessons to teach, as these lines I adapted from 'The Merchant of Venice' doth attest," Henry responded.

"I feel that I ought to cook something for them," Sarah suggested. "I can make some fried pies, something like a Boreka Tart or Sopapilla with those guavas. When will they come?"

"When the boys have finished their lessons," Aron stated. "Sometimes this afternoon, I think."

"Fred, can I talk to you about something?" Aron asked.

"Sure," Fred replied.

"Let's go out into the hall. Alex, would you come too. We have some man things to talk about."

The three men went out into the hall and down to the next landing.

"Fellows I have a problem that I can't talk to my Dad or Mom about," Aron started. "We've got to use condoms since we found out about our chromosome problems which could result in our baby having a fatal liver complication. This is called Haemochromatosis. Sam could not get enough during a part of her pregnancy, but now she has cut me off."

"That's not your child?" Alex asked.

"We had in-vitro fertilization and implantation after the egg and sperm were tested to make sure they were all right. From all indications everything worked," Aron answered. "My sister's kids are all right because her husband did not carry that defect. It is just that Sam and I had the bad luck to both have it."

"I'm sorry, but what does this have to do with us?" Alex remarked.

"It's him," Alex responded pointing to a rising bulge in his pants.

"You know what to do about that," Fred responded.

"I can't," Aron explained. "It has been pounded in my head ever since I was a child that men are supposed to save themselves for their wives. I look at Sam and I want her, and up he comes."

"There are a lot of ladies in Detroit that can take care of that for a few dollars," Alex suggested.

"And then I come home with syphilis, herpes, AIDS or this new Monkey Pox?" Aron pondered. "I can't do that to Sam."

"You have wets at night sometimes, don't you?" Fred asked.

"Yes, but when I do I feel guilty about it, particularly if the dream has someone else in it other than Sam," Aron admitted.

"There are other ways to take care of that sort of thing, but working that out is between you and your wife," Fred concluded. "During the sixties everyone was having sex with everybody else. When I met Lollie I quit doing the drugs and sex thing. We worked things out, and you and Sam can too."

"Thanks guys," Aron responded as he pointed to the growing bulge in his pants. "But that did not really help."

"Go to the bathroom and take care of that," Fred suggested. "It's okay. You won't go blind."

When the three men reenter the living room Aron is walking rapidly towards the bedroom bath with a bulge in the front of his trousers.

"What did you men talk about?" Sarah asked.

Lollie pointed to the front of Aron's trousers as he increased his pace and shut the bedroom door behind him.

"I see," Sarah concluded. "That is what men always talk about."

"Is there some problem between you and Aron?" David asked.

"He wants it, and the whole idea disgusts me right now. We'll work it out.

Chapter 20

Sextuplets

"I want to try the baby on the bottles and the milk and see if he will take them," Samantha stated to change the subject. "Let's get the formula made up and the bottle ready."

"It has been awhile, but I remember how to bottle feed," Sarah replied.

"Me too," Lollie agreed. "It would be good to get him use to the bottle and formula while we are all here. Thank God for the microwave. We can use it to sterilize the bottle and heat the formula, rather than boiling everything."

Aron answered a knock on the door. The Pasha family, all wearing masks, entered; and Aron motioned for them to sit on the couch.

"This is my wife, Samantha," Aron began. "My mother Sarah, my father David, Sam's mother Lollie and her father Fred. These two gentlemen are our movers, Alex and Henry who are snowed in with us."

Erderlan likewise introduced his family, "My wife, Keje, and my sons Bryan and Tom."

Bryan and Tom have never had such a close encounter with a black man and stared intently at Henry.

"It is not polite to stare at people," Keje reminded her sons.

"Greetings young squires," Henry replied. "It is good that you gaze so fixedly at someone so different from yourselves. My appearance startles, my voice be strange and my dress unfamiliar. But such in your travels you have seen many times before. Come, recount your tales of what witnessed you have."

"You talk strange," Bryan observed.

"Tis a Shakespearean actor I wish to be," Henry responded, "and this voice I effectuate is to that aim intended. Like your Persian poets of old that are not in my knowing, he in English wrote of things wondrous, strange and beauteous."

"Please." Erderlan requested. "Speak simply to the children. They are trying to learn standard English, and we do not want to confuse them."

"Understood," Henry responded. "I but....I understand. I'll try."

Keje went to the bassinet, looked at the sleeping baby and remarked, "That is a beautiful child. How old is it?"

"We just brought him home from the hospital," Samantha replied. "He is going to wake up soon and need feeding."

Aron responded to that comment by taking the bassinet into the bedroom and Samantha followed.

"I was thinking of making some borekas using those canned guavas that you gave us," Sarah suggested. "Would you like to help?"

"We call them bourkas, but I suspect they are the same," Keje answered. "I would be glad to help."

Henry and the boys went into one corner of the room. Keje, Sarah and Lollie went into the kitchen. By this time Aron had returned and joined Erderlan, David, Fred and Alex on the sofa.

"I know what we just went through moving into a new house with a new baby during a snowstorm," Aron began. "I can't imagine what it was like trying to escape from a war zone with your wife and boys."

"We never knew from one day to the next whether we were going to eat, be shot at, freeze to death or be put into some internment camp for who knows how long," Erderlan began. "We managed to get away from Isis and into the Kurdish part of Syria, and then to Turkey and here. Many, many were not so lucky. I helped treat people who were starving, had frozen feet, diseases of all sorts and wounds from mines and gunshots. Because I was young and had medical training, we were given preferential treatment."

"How far along were you in your studies?" David asked.

"My third year. I was supposed to intern in a hospital before taking my exams, but the war interrupted all of that."

"Does that mean that you have to start all over?" Aron inquired.

"Without my records, I can't get credit for what I did or afford to go to med school," Erderlan replied with regret. "Maybe I can be a nurse or EMT, but not a doctor. But even that seems out of reach at the moment."

"With the hospitals filling up with patents with this virus, one would think that they would be happy to have anyone with medical training," David observed.

"While that is true," Erderlan agreed. "I still need certification. Our aid worker is trying to link me up with a teaching hospital. However, she is swamped with refuges who have even more pressing problems."

"I work with the University Hospital," Aron suggested. "I think that I may be able to get you into an orderly's position and then training as a respiratory therapist. Then maybe you can complete your studies."

"Anything along that line would be helpful," Erderlan agreed.

The group's conversation was interrupted when scratching and whimpering was heard from the closet where Momma Dog was starting to have her pups.

"I think that someone may be in need of your services right now," Fred observed.

Henry, Bryan and Tom were sitting on the floor playing a word game.

"What has four legs, stripes and hunts at night?" Henry asked.

"A tiger," Bryan replied.

"Tom, where are tigers from?" Henry queried.

"India," Tom replied.

"Bryan a harder question," Henry prompted. "What part of India are they named for?"

"Bengual?," Bryan answered with a hint of doubt in his voice.

"That's nearly right." Henry complemented. "Bengal."

"Tom," Henry continued. "Where else are there tigers?"

"Suburbia?" Tom responded.

"Not very often I hope," Henry answered with a smile. "But you are so close. They are also in Siberia."

Erderlan walked over to the group and informed them, "The dog is having her pups. Come see."

"The first one is coming out," Bryan observed excitedly. "What do we do?"

"Nothing right now," Fred replied. "Dogs have been having pups for millions of years and at this stage are best left alone. Only if the pup is coming out sideways do we need to do anything."

"How many will she have?" Tom asked.

"There is no way of knowing," Fred responded. "There might be six or more. It is going to take some hours before she is done."

"I'll go tell Sam," Aron remarked, "although I don't know if she wants to see this after just having a baby herself."

"What's she doing?" Tom asked.

"She is washing up the first pup and will eat the afterbirth and cut the umbilical cord with her teeth," his father replied.

"She's doing it," Bryan observed. "Do human moms do that?"

"Thankfully not," Sarah replied having walked over from the kitchen.

Bryan reached down to touch the pup, but was stopped by his father. "Don't touch it. Her washing it is putting her scent on it. The pup is blind and will nudge its way to a teat and start nursing. If it has a problem, lift the blanket and move it in the right direction."

"Samantha said that the birth experience was too fresh for her," Aron said as he rejoined the group. "She was afraid that seeing this would trigger sympathetic birth pains."

"There is something to that," Sarah observed. "Sometimes when one woman gives birth it will trigger the same response in others."

"The next one is coming," Bryan observed. "It's white. Are they all going to be different?"

"They may be," Erderlan answered. "If they have different fathers, they will be different colors."

"Just like us," Alex observed. "Things seem to be doing okay here. I think that we should leave Momma Dog alone until she gets done. It's getting about bedtime, and we are going to have to try to dig out tomorrow."

"I'll look after her for a while," Henry suggested.

"Can we come back in the morning?" Bryan asked.

"We can come down after you have done your math," Erderlan responded. "Will that be all right?" he asked Aron.

"I don't see why not," Aron agreed. "We are not going anywhere."

"I can't wait," Tom remarked enthusiastically.

Henry kept watch while Momma Dog delivered six pups who were now sleeping or nursing. Satisfied that the last had been delivered, he walked over to the day bed, pulled up the covers and climbed in. as he slept, he dreamed. This time he was pulling a huge stone up a dirt path while others were putting rollers under it and a Taskmaster lashed out with his whip.

"Pull you scum. Pull," the Taskmaster ordered.

Henry pulled but lost his footing as the stone pulled him down and dragged him back down the hill. The Taskmaster sat him upright and reattached him to the block. Henry looked up the incline into the blazing sun to start his task again. The whip beat across his back, and he woke with a start in a disheveled bed in a sweat. He then pulled the bedding off the daybed onto the sectional couch and sleepily covered himself and lay down again.

"Don't want to go back there again," Henry thought to himself.

"Up Droopy Drawers." Alex ordered. "It's morning. The snow's quit, and we have to dig out today. What are you doing over here?"

"Sleep terrors had I in desert Egypt pulling blocks of stone up a hill only to have them slide back again. Then start over would I. Repeat, repeat and repeat again. In the land of dreams, slept I not."

"Get that stuff off the couch and make that bed up again. Everyone else will be up soon." Alex instructed. "How did Momma Dog do?"

"She had six when last I saw," Henry replied.

Aron, Alex, Henry, David, Fred and Samantha sat around the table with their coffee while Lollie and Sarah were washing dishes in the sink.

"Thank goodness the snow has stopped," Fred observed. "That is more snow than I have ever seen."

"That was a bad storm, even for around here," David agreed. "That is more snow than we have had in more than 20 years, I think. Has anyone seen a snow blower or a shovel?"

"I've got a shovel in the truck," Alex responded.

"There was another inside the laundry room," Aron reported.

"I've got to get that snow off my truck and move it to break the wheels out," Alex outlined. "Then when they have a lane plowed, we can leave."

"That dog is going to have to go out or we are going to have a mess in here," David observed.

"To that task I will myself apply," Henry responded. "She and me and they have the dark hours spent in intimate company as midwife I played. With plastic bags and gloves that task I will gladly do as soon as we have a pathway shoveled."

"Droopy Drawers and I will get started," Alex suggested. "You other men can relieve us in a half hour and we can swap off until we get finished."

After Alex and Henry left, Lollie walked over to Samantha and asked, "How are you and young Michael this morning?"

"Reasonable, I guess," Samantha replied. "He got us up every few hours because he was hungry or needed changing. Aron and I are both really tired."

"Sarah and I sterilized a bottle and made up some formula," Lollie reported. "Do you want us to heat it up and try that on him? We'll feed him and give you some rest."

A few minutes later the exchange had been made and the baby nuzzled the nipple and then took it. "Good, Sara reported. "It looks like he is not going to have any problems with that."

While Lollie fed Michael, Sarah walked over to Momma Dog and asked, "How is our other mother today?"

Momma Dog lifted her head in acknowledgement, wagged her tail and licked Sarah's hand to express her gratitude.

Outside Alex shoveled out from around the wheels of the van while Henry was digging out the garage door. Henry bumped the door, and a cascade of snow plunged off the roof and nearly buried him in a cold, white mound.

"Droopy Drawers," Alex kidded. "You are supposed to be shoveling snow away, not adding to it."

"My task sir, I know," Henry replied. "But this snow doth blow and immersed am I in this chilling blanket of white."

Henry pulled himself out of the snow and recovered his shovel. He then took a shovel full and threw it off to the side only to have the loose powder flow back into the hole. Ultimately, he managed to make some headway and shoveled a pathway to the street. Alex had somewhat better luck with a broom sweeping snow from the top of the van. When Henry approached, he receives a fresh splat of snow as a slab came off the van.

"Retire I will to Momma Dog retrieve, and remove thy newly deposited presents," Henry stated as he shook the snow off his clothes.

"While you are up there, ask someone else to come down," Alex instructed.

As Henry, Fred and Momma Dog attempted to descend in the elevator, Henry punched the down button several times and nothing happened. Then on the next try it dropped several feet caught and then discharged its passengers at the garage.

"Although no spirits now hear I speak," Henry observed. "This machine appears possessed by some evil forces."

Henry and Fred arrived in the garage and found Alex eyeing the bottom of the garage door which was frozen to the pavement.

"When we go outside, I am going to try to break the garage door free with my shovel," Alex postulated. "While I try Henry take your shovel and see if you can knock it lose anywhere it is sticking."

With Momma Dog free to run in the cleared spaces, Henry put her down and attempted to drive his shovel between the pavement and the bottom of the garage door. The garage door motor ran, but the door did not move. Sliding the shovel along the base of the door, it broke free, lifted and another cascade of snow came off the roof which buried Henry under a mountain of white. Alex and Fred pulled him out. Soon after, Momma Dog returned and whined to go back inside.

"Take her up," Alex told Henry. "We'll finish up here. Until the snowplow comes, there is not much we can do."

A knock on the door announced the return of the Pasha family who were let in by Aron. They, along with Henry, immediately gathered around Momma Dog and her pups. Bryan and Tom spoke in Kurdish to their parents. There was no need for translation as it was obvious that the boys wanted one of the pups.

Henry responded, "With their mother they need to stay until eat, and walk and see can they – for a full moon think I. Mrs. David will keep until mobility they have gained. My purse and lodging are too light and small for more than Momma Dog to keep."

"Can we have one please," Bryan pleaded. "Tom and I will take care of it. You won't have to do anything."

"Taking care of a puppy is like adding another member of the family," Edelan informed. "There are doctor bills, food to buy and some toys. What I must do to provide for it takes away from what I can give to you. They are city dogs, and they have to be trained – just like you go to school."

Bryan looked at Tom and he nodded in agreement.

"We'll do it," Tom promised. "It'll be like another brother – only with fur and four legs."

"What does your mother have to say?" Erderlan asked Keji.

"We could never have pets as children, and I always wanted one," Keje said. "I think that caring for it would be a good learning experience for the children as there are no sheep, goats or horses in their lives here in the city."

"I am going to take them home and look after them until they are old enough," Sarah responded. "Which one do you want?"

The Pasha family had a discussion and settled on one of the larger long-haired female pups.

"What are you going to name it?" Keji asked. "It is an American dog, so it should have an American name."

"Marilyn, because it has blond hair," Tom suggested.

"What about Rover or Fang?" Bryan suggested.

"I can have a Rover in the house," Erderlan agreed. All agreed? Keje. You too?"

Bryan and Tom happily agreed and Keje sealed the decision with the statement, "Then Rover will be a new member of our family."

After the men had worked all morning the garage door and area around it was shoveled clean. Snow from the high piles on either side of the driveway was melting and water dripped from the sides of the van.

"I need to shovel out the the car that I drive," Erderlan said. "It's two doors down."

"I'll go and help so you can drive everyone home," Aron volunteered.

Aron and Erderlan took shovels and postholed down the sidewalk while Alex, Fred and David watched them walk away.

"What's that noise?" Fred asked.

As he spoke another slab of snow broke free from the roof peak and with a rumble broke over the men as they attempted to cover themselves with their arms.

"Damn," Alex responded. "That's cold. Is everyone okay? There is nothing more we can do here until the snowplow comes. Let's go inside."

While everyone was sitting on the sectional sofa Lollie was gazing out the window.

Momma Dog went and stood on the couch. When the dog began

barking as she looked out the window. Lollie walked over and announced, "The snowplow is coming. I see it down the street."

Everyone went to the window and the men watched in anguish as their clean driveway and moving van are once more covered with snow.

"I guess they had plows, but no trucks," Alex said. "We'll have to do it all over again. Come on Droopy Drawers, we've got to get out of here."

"For labor I wanted and labor I got," Henry observed. "But yea, full more and double the friendships here made by those who fair chance bound together with icy bonds. I bid you farewell with affection and memories that time will not soon erase bound yet more firmly by the spawn of recent births."

An hour later goodbyes and hugs are exchanged as Alex and Henry prepared to get into the van. David and Sarah are loading Momma Dog's bedding into Erderlan's car to take them home as Fred and Lollie prepare to walk to their bed and breakfast.

These departures are watched by Momma Dog who stood on the daybed looking out the apartment window. As she stood on her rear legs with her front paws over the back of the daybed her claws ripped the fragile cloth on the sides of the bed. A trickle and then a rush of gold coins poured from the bed onto the floor.

Chapter 21

Discovery

Aron who was carrying a laundry basket to take the puppies downstairs, opened his apartment door and his attention was immediately drawn to Momma Dog who turned around on the daybed and released more coins from the bed which joined the growing pile on the carpet.

"Look," Aron whispered to Sam. "I don't believe it. No wonder that thing was so heavy."

The pair ran over to the daybed and Aron ran his hand beneath the ripped fabric and pulled out more coins. Then he tore the cloth from that portion of the daybed exposing rows of coins set in decomposing glue and horsehair.

"There are hundreds of them," Samantha said. "Hundreds. Aron, we are rich. Be careful, who knows what germs are in there after all these years."

This warning sobered Aron somewhat as he started not only to think about ridding himself of potential infectious agents, but what his dad might say if he knew about this unexpected family treasure that he had given away.

"We've got to be calm and take the puppies and Momma Dog down just like nothing had happened," Aron said. "We need to see what we've got and decide what to do with it. Dad and Mom should certainly have some of them, but we have to work all of that out later."

As Aron put the shipping blanket and the puppies in the basket, Samantha went over and took Momma Dog into her arms and hugged her. "You did good, Momma Dog. I know you don't know what you did, but you and yours are going to be very well taken care of."

With the wiggling puppies in the laundry basket and Momma Dog following them into the elevator, Aron and Samantha returned to the garage where Erderlan waited with David and Lollie in the car.

"What took you so long," David demanded of his son.

"It took us a few minutes to say goodbye to Momma Dog and gather up things," Aron said as he helped stow the basket in the back seat of the car and Samantha gave Lollie the Crock Pot. With a round of kisses and hugs from Lollie, Erderlan started off and Aron and Samantha watched them drive away.

Not waiting for the elevator, the pair ran back up the stairs and prepared to examine their new-found wealth. In preparation they got pairs of rubber gloves, cut out the bottoms of some of the boxes to make flats and got some old tooth brushes to clean the coins.

As they brushed off the coins, they separated them into stacks of similar sizes and countries of origin. By the time they are done there were more than 1,500 gold coins were arrayed in stacks in the bottom of the cardboard flats.

"Most are French, but others are from all over Europe with a few from the U.S. and South America," Aron observed. "Some are worn from circulation, but others are almost new. I think that we are going to have to sell these say maybe 20 at the time through a coin dealer that we can trust to get the most out of them."

The telephone rang and Samantha answered. "Aron, it's your sister. She says that she has been trying for days and finally got through. I'll put her on speaker."

"Sam, Aron, this is Beth. I have been trying to reach you ever since I heard about the baby. How are you all doing?"

"We are getting along fine," Samantha answered. "The baby,

Michael, is now around eight pounds and is growing like a weed. He is going to be a big guy like his dad."

"Aron," Beth said. "I owe you an apology for walking in on you in the shower and throwing your suitcase out of the window. Things with my kids at school and with Larry Jr. had reached a boiling point, and I was mad at the whole world at that moment. I needed some time to put things back together."

"Your stepdad explained," Aron replied, "and it is all right. We had more adventures, this time with Georgia ticks, after we left."

"So I heard from Elizabeth and Margaret," Beth agreed. "They got quite an education picking ticks off you. I am worried that everyone saw rather more of you than you expected."

This statement triggered a memory of him lying naked on towels in a large room with having ticks picked off him by 18-year-old Margaret who was dressed in shorts and a tank top while Samantha was being picked over by 20-year-old Elizabeth.

"That was quite an experience," Aron agreed. "How are things doing with you and Larry Jr.?"

"Larry Jr. and I both wanted to come up when the baby was born, but since we married, he has been working on that huge warehouse in South Dakota. He has had all kinds of trouble getting electrical supplies, and while they are pouring concrete he had to be there. I have my six educationally challenged kids to look after and could not get away either."

"I have never met Larry Jr., but I am told that I would probably like him," Aron mused.

"He said that he would like to talk to you about doing something together after he gets through with this job," Beth responded.

"How are his parents doing?" Aron asked. "I hear they got really sick after his college roommate puked all over his father at our wedding."

"Both Larry Sr. and his wife were taken to the Dallas hospital as soon as they got back to Texas. He had to be put on ventilation, but came through that. Interestingly, the guy who puked on him, Albert, is in West Texas on their ranch taking care of them. They are doing to take some time before they are really well."

"I'm glad to hear that," Aron said. "What does Larry Jr. want to do?"

"When you were down at Sam's father's place, he made you an offer of a week-long survival experience offered by a bible camp and nudist colony in Terlingua, Texas," Beth explained. "Larry wants you and him to go together and spend a week there."

"What would they do?" Samantha asked.

"This is a 'Naked and Alone' type of experience," Beth continued. "Larry Jr. has lived in this country most of his life and says that he will take Aron through it. He said that this will build his self-esteem and earn him some creds with his dad."

"It is true that Dad has always considered me a 'Momma's Boy,' and beaten that into me all of my life," Aron elucidated. "I think that I would like to try something like that."

"Aron, not to get too personal about it," Beth advised, "Before you start running around without any clothes on, you had better get a very dark tan. You are as white as a naked mole rat. That is something that you need to start working on now or you will blister like a fried pork skin."

"I'll have to ask Sam about the mole rat business," Aron chuckled, "but you are right about the tanning. They have tanning beds at the gym, and I can start with that."

It was almost all they could do to keep a since of composure over their new-found wealth as that pile of gold on the floor drew their interests like a super magnet drawing them to it.

"Running her hands through the coins," Samantha asked, "How much is all this worth?"

"I don't know and no one else will know either until they have been cleaned, sorted, graded and sold," Aron responded. "Gold is high right now, but except for coins that are too worn to be of interest to collectors, they are worth more than the bulk price for gold. They are all alloyed, so they are not pure gold either. We will have to sort them, count them out and then see what we have."

With toothbrushes and alcohol, Aron sat on a sheet cleaning and separating the coins by type and nationality. "What I am seeing is that most of these are French from the time of Napolean II and a scattering of others from Europe and the U.S."

"The most recent date that I see on any of them is 1870," Samantha observed. "Let me get my computer and see what was going on in France at the time. Maybe that will give us a clue."

"There is more demand for U.S. Gold coins in this country than any other," Aron told Samantha. "There are some $20 Liberty Heads and $10 Indian Heads. These we will have no problem selling at all. The English gold Sovereigns will be no problem either. I suppose that there is an international market for the French coins, but I think that it will be harder to sell them here. We are going to have to find a dealer that we can trust to handle this for us."

"It looks like these coins were in circulation at the time of the Franco-Prussian war when the Germans invaded France and marched to the outskirts of Paris," Samantha reported. "Someone apparently put these coins in this piece of furniture to ship some of their family fortune out of the country."

""That fits with what I have always been told about that day bed," Aron said. "It was built by one of my ancestors, taken to England, and apparently after none of the French survived the war, he brought it to America. The gold was somehow kept a secret all this time. I suppose that the man who built it died before he could pass on that information to his children. The family passed it on through the generations out of a since of duty, but none of them knew of the gold inside."

"What are we going to do with it?" Samantha asked.

"I think that perhaps the first thing we need to do is to go online and buy some plastic coin tubes to store them in and then put them into a safety-deposit box in a bank," David suggested.

"What we don't want to do is to attempt to sell them all at once," Samantha responded. "That would attract the attention of the IRS and they would want taxes on them. If the word got out about this, we could be sued about them with everyone wanting a share, whether they had any claim to them or not."

After Aron had finished his tally he concluded, "There are a few over 1,500 of them. If the average weight is one-half ounce that means that they are worth something like $750,000."

"That is a lot of money," Samantha observed. "But we will have to be careful with it or we can blow it all in a hurry. We also need to share it with your mom and dad, Henry, Alex and the Pasha family upstairs. They all participated in this adventure we just had, and it was, after all, Momma Dog who discovered the coins."

"Lets' say we split 75:25 between us and the rest," Alex suggested. "As we sell off the coins over the next ten years everyone would get a check. The dealer will take ten percent, and I think all should be satisfied."

Nathan's Coins and Stamps was a store front in a strip mall located at the site of a former jewelry store. This was a nearly ideal location for Nathan Rainwater's operation as it was already equipped with a vault and advanced security system. It had been a somewhat slow day when Aron and Samantha walked in.

"We inherited some gold coins that we would like to sell," Aron said.

"What do you have?" Nathan replied.

Aron pulled out a plastic tube which contained a mix of worn and battered coins that he had selected.

Sam took a loop and examined each one, weighed it, and kept a running tally on a pad. He then went to his computer to check the current prices on the items and returned to the couple.

"These are an interesting assortment, but are too worn to have much collector's value," Nathan reported. "This is what I can offer for their scrap gold value."

He passed a note to the couple who stepped to one side.

"Is this about what you expected?" Samantha asked.

"Considering that these are the worst that we have, I think that this is about right," Aron replied.

"Nathan," Aron began. "We have a large number of coins in better condition than these to dispose of over the next decade. We need someone to grade them, sell them, and send checks to us and four other parties. Can you take care of this for us?"

"I take it you want to get the best prices for your coins and avoid the IRS?" Nathan responded. "How many are we talking about?"

"Over a thousand," Aron informed.

"Wow," Nathan interjected. "I have had people come in with ten or twenty, but nothing of this magnitude. I can set up a special account and cut the checks against that. I would charge a broker's fee of 10 percent on the transaction. Would that be satisfactory?"

"Over the next months we will try you with three batches of 20 coins each," Aron said. "After that I will bring you more from time to time. Here are the addresses where I need the checks to be sent and the percentage splits. If all goes well, this relationship can continue for years."

"Dad, Sam and I want to thank you for that day bed that you gave us," Aron began as he talked to his father on the telephone the next day. "We figured out why it was so important that it be passed on through the family."

"That was something that I did not have any attachment to, outside of the stories that I had always been told about it." David replied. "What new information do you have?"

"The reason that bed was so heavy was that it was stuffed full of gold coins from the 1870s – hundreds of them."

"Hundreds you say," David replied incredulously.

"Yes hundreds," Aron responded. "They date from the time of the Franco-Prussian war which was the time that … who was it, Issac? left France with his wife."

"Sam and I have decided to sell them off in small batches and keep some of the money and distribute some of it to you and Mom and to Henry, Alex and the Pasha family upstairs," Alex apprised his dad. "As they are sold you and mom will be sent a check from the dealer."

"How do you know you can trust him?" David inquired. "That is a lot of money."

"We are only going to give him one batch at the time, so it is to his advantage to get the best prices he can for them and treat us honestly so that we continue to deal with him. Most of the coins are French and are a little more difficult to sell in this country than they would be in France. However, he has access to the international market and can ship them anywhere in the world."

"I would like to have some samples, so I could at least see them," David requested.

"I will put together sets of the best of them and frame them – one for you and one for Rose. These are part of our family's legacy, and I feel that we all should have some of them. They will be this year's Hanukah presents," Aron answered.

"While I cannot deny that Lollie and I could always use some extra money to maybe travel a little. Are you sure you are keeping enough for yourself, Samantha and Michael?" David asked. "Even a million dollars does not go nearly as far as it once did."

"We are going to live just as we did," Aron replied. "That extra money is going to be a cushion for hard times and maybe to use if Michael has any developmental problems. As long as I stay with the hospital our insurance costs are covered, we can pay off the house earlier and get another car sooner than maybe we could otherwise. We are not going to blow it all on stuff we don't really need."

"That's prudent," David agreed. "You and Sam have apparently thought this out fairly well. I will save this surprise for your mother."

"There is one other thing Dad," Aron continued. "It was Momma Dog who found the coins. She jumped up on the daybed to watch everyone leave and when she did she ripped the cover of the daybed and released the coins. We want to make very sure that her and her pups are very well taken care of."

"I am not enthusiastic about dogs as you know," David replied. "Sarah has taken very good care of Momma Dog and her pups, and they have all made their first trip to the vet. That dog has been treated and fed better than many people are today. As long as she is under Sarah's care there will be absolutely no problems – no problems at all. I might not eat, but that dog will."

Getting Alex's and Henry's telephone numbers and addresses were somewhat more of a problem, but it was eased after Samantha assured Alphonso of Swift Movers that she was not trying to hire Alex and Henry away from him.

"Henry, this is Samantha," she informed. "We have received an unexpected inheritance that I cannot tell you about, but we would like to share some of it with you and Momma Dog."

"What say you kind lady," Henry replied. "For scratch I am lean of, and money is an all-time concern, particularly as another family member I will be attaching to drain my meager resources."

"From time to time you will be receiving a check in the mail as we receive the money," Samantha explained. "The checks will vary in amounts. They are intended to help support you and Momma Dog and help you further your career. What's important is that you keep your address current so that the checks can reach you."

"Move I will, and must as opportunities break over distant horizons, but keep in touch I will," Henry assured. "Acting comes in spits, spirts, rushes and lulls; and never I know what might lie ahead. Any income would be a considerable aid in keeping me afloat and Momma Dog in vittles. No questions I ask, but forever thanks for your kind generosity."

"Alex, this is Aron. You moved us a few days ago and we have received an unexpected inheritance that we would like to share with you."

"Is this some kind of joke?" Alex asked. "You don't know me from beans. Why would you want to give me money?"

"I cannot give you the details, but we unexpectedly received an inheritance that we will be selling off in coming years and would like to give you a cut of it," Alex assured the doubting teamster.

"No one gives something for nothing, what do you expect from me?" Alex insisted.

"Alex, think of it as if you had won the lottery," Alex assured. "Periodically you will receive a check. You may do whatever you want to do with the money. It's yours – all yours. I can't say how much it will be or how long it will last, but this will be paid out over a period of years."

"Well thanks, I guess," Alex responded. "I never have accumulated much in the way of savings and some extra seed money would be very useful at this time in my life. I will look forward to getting the checks."

"The only thing to be careful about it," Alex warned, "is that we have your current address and you have a safe place to get the checks. They can be sent to a bank if that is what you want."

"Erderlan, this is Samantha. Can we talk?"

"I am at home now. Do you want me to come down?" he asked.

"That would probably be best. Aron has arranged a job interview at the hospital for you, and we have some good news for you as well."

When he arrived, he sat down on the sofa with Samantha and Aron.

"Things are a madhouse at the hospital right now," Aron began. "I have set up an appointment with the personnel officer at 10:00 AM tomorrow morning, and I will go with you. They are moving and shuffling around offices and beds, and I will help you find him. You will need to bring your IDs and social security card to get signed up. They will also require a physical which may or may not be able to be done today. I just don't know."

"That is wonderful," Erderlan replied. "Without some sort of 'in' I could not imagine how to even go about applying for a job."

"We have some other news too," Samantha added. "We have received an unexpected inheritance that we will be selling off over the next decade or so. Aron and I would like to give some of that to you and your family to help you get through medical school and become the doctor you always wanted to be."

"I'm...flabbergasted or whatever the word is," Erderlan replied. "I don't know what to say. This is beyond anything that I ever dreamed might happen."

"The money will come to you in checks at irregular intervals and amounts as the property is sold over the next five or ten years," Samantha explained. "Just let it accumulate in a bank and draw from it as you need. It will build up to a very significant amount over time. It will not be enough to buy a house, for example, but it should be enough to help cover your med school costs and reduce the amount you might have to borrow or get grants to cover."

"I am so very grateful for what you have offered to do and done," Erderlan replied. "This is exactly the sort of 'leg-up' that I needed."

THE *Goldfarb* CHRONICLES

Book 2

The Solitario

Chapter 1

New Horizons

LARRY HAD NO PROBLEM SPOTTING Aron as he collected his bags at the El Paso airport. He stood noticeably taller than the other passengers, now had a tan and was dressed in a sports coat, open-collared shirt and slacks compared to most of his military fatigue or jean clad fellow passengers.

"Aron, over here," Larry shouted from the other side of the exit turnstile.

Aron grabbed his suitcase, roll-on and computer bag and after having the attendant check his tickets, joined Larry in front of the exit doors.

Wanting to get started on their 250-mile trip to the ranch as quickly as possible, Larry said, "I've got one of the ranch trucks outside. We'll have a chance to talk on the way. I've heard so much about you that I feel I already know you and am glad to finally meet you."

"Me too," Aron responded. "I have been looking forward to meeting you as well."

"We will stop and eat along the way," Larry said. "Do you need a bathroom before we get started?"

"No," Aron replied. "I already took care of that. I'm ready to get out of here. I've seen enough of airports for awhile."

"I bet you have," Larry thought knowing that Aron has flown from

Detroit to Chicago to Dallas and finally to El Paso. West Texas was a part of the world that was not easily accessed from hardly anywhere, which was to him part of its greatest attraction. He wondered if his city-boy guest would come to feel the same.

"Is everything going well with Samantha and the baby?" Larry asked once they had cleared the airport traffic and were working their way out of El Paso into the vastness of Hudspeth County where there were fewer vehicles on the interstate as the dominant populations changed from humans to cows, horses, antelope and deer.

"We had a rough time with the baby and moving into our new house, but things are much more settled at present," Aron reported. "Sam is doing less flying and more remote work from home which is helpful, and I am also mostly working from home. My parents can sometimes take the baby, which really helps out. How are things doing with you and Beth?"

"Our situation is much like yours," Larry said. "Beth has been almost completely absorbed with her disabled kids in the school system, and I have been away on that warehouse job in South Dakota. That was good work and paid well, but I am glad to have finished with it. With what I hear is that they have installed an automated system that is supposed to handle the sorting and shipping of thousands or even tens of thousands of orders a day, but are still having problems with it. My part of it, designing the electrical system and getting the necessary voltage in the right places, is thankfully 'set in stone' so to speak, and I am out of that mess – and glad to be. With things constantly breaking down and computer systems failing and trying to meet shipping deadlines is driving the poor site manager crazy. In addition, some of his subordinates are trying to stab him in the back by blaming him for everything that is going wrong with a system that was not sufficiently tested before taking it to large-scale applications. I am so very glad to be out of that mess. Ultimately, they will work everything out, I suppose, but that is thankfully all their problem now. I am completely out of it. I'll probably get similar jobs for other large distribution centers, but I will be happy to be done with this one."

"Wow. So, I have been hearing from Sam," Aron expressed

sympathetically. "I can see why you would want to get away from all of that. Tell me about this country we are driving through."

"Aron, I'd be happy to," Larry replied. "My family and I love this unforgiving country, and I have been looking forward to teaching you about it ever since I heard that you were going to be marrying Sam. I can't say exactly why. Maybe it is because we have had such different experiences and opinions that if I can show just one person how I feel, the world will be a more understanding and less confrontational in-your-face-and–fuck-you type of place."

The large crew-cab Dodge truck allowed Aron to adjust his seat to the point where he could extend his legs to where they were more comfortably situated than they were in the airplane cabin. "At least," he thought, "I've got interesting company and something better to look at than the back of another seat. It also feels a damn site better to get that mask off."

"We humans have short lives and feel like things should never change," Larry began. "What we don't appreciate is that climate change has always been with us through forces that we cannot control. Only about 10,000 years ago this was a different place when man first came into this country.

"You have seen what glaciers did in New York state as they carved deep valleys and cirque lakes. What the Paleo-Indians saw here were lush green well-watered valleys and plains filled with giant species of wildlife like mammoths, elephants, super bison and saber-toothed tigers which they followed and hunted.

"The rain that fell then is now trapped in our groundwater and is pumped to water our crops and cattle today. Ever since that time the climate has been generally getting dryer and dryer and the further south we go towards the ranch the drier it gets. In the higher areas like around Alpine and Marfa the ranchers grade their land on how many cows can be supported on an acre of land. Down here on my dad's ranch is it is how many acres it takes to support a single cow. The native desert grasses are more nutritious, but we also grow crops like alfalfa in valleys where we can get water to it.

"We started out in the block faulted mountains of the Great Basin and as we get towards Alpine and Marfa we will progress to areas where volcanos and igneous activities as well as tectonic plate movements have impacted the landscape. The place where we will be put out The Solitario, The Lonely One, is the remains of a huge volcano that blew its top and scattered ash in layers that were from 20 to 100 feet thick over hundreds of square miles. There was a lead mine in the center of it, but was never a commercial success."

"What does anyone do out here?" Aron questioned.

"In the larger towns you have the usual fast food and chain stores that have run most of the independent merchants out of business. There are still some local restaurants, lawyers, barbers, doctors, dentists and so on. These support local farmers and ranchers. There was once a fair amount of mining going on, like the mercury mines at Terlingua, but not much is happening now except for sand, gravel and cement making. When you get down to Study Butte where my dad's ranch is you have one store-post office that carries everything from gas to aircraft parts. They say, 'if we ain't got it, we can get it,' and they do."

"Larry, I see we are coming up on Van Horn," Aron said. "Can we stop and get something to eat. I am thirsty as hell because of the dry air in the airplane and here."

"We can," Larry replied. "I need to stop anyway and see if I can pick up some whisky for Dad. He is partial to George Dickel Rye Whisky when he can get it. They don't have it all the time, so he likes for me to check when I pass through. Is there anything that you want me to pick up for you?"

"A few cases of Bud Light will do. I feel like I could drink a whole one right now," Aron kidded.

"We have mostly Lone Star at the ranch," Larry replied. "If you run out of Bud, Lone Star is a similar light beer. I will get you some Bud while we are here. I'm sure that it will not go to waste. I'll cook javalina in it if nothing else."

The Longhorn Café was two blocks south of the tourist strip on the main street of town and the line of dusty mostly beat-up pick-ups with horse and equipment trailers parked in its lot attested that this was a place preferred by the locals over the fast-food joints.

"Sit anywhere you like," the waitress motioned as they walked in. "We've got some really good bar-b-que ribs that are just out of the cooker and a variety of steaks and side dishes," she said.

"What do you feel like eating," Larry asked.

"A tall cold frosted beer, a glass of ice water, some salad, a baked potato and some of those ribs sounds good," Aron said hungrily. "My stomach feels like it is still at 30,000 feet somewhere."

When the waitress returned, Larry ordered. "Please bring us a tossed salad with thousand island dressing, the rib platter for two, a baked potato, beans, a large cold bud light and ice water for him and unsweetened iced tea for me."

"I hope that is all right with you?" Larry asked before the waitress left.

"That will be fine Larry," Aron replied. "Pointing to a huge stuffed longhorn cow head with six-foot horns over the fireplace, Aron asked, "Is that the kind of cows that you raise on the ranch?"

"We do have some," Larry replied. "But mostly what we have are Brahma crosses that can tolerate the heat better than the usual Angus beef cattle. We also have Angora goats that we grow for their wool. Dad was always interested in having a variety of animals on his ranch."

The waitress brought their drinks along with large bibs for the men to tie over their clothing. When Larry detected a questioning look from Aron, he replied, "There is almost no way that you can eat these without getting the sauce all over you. You break them apart with the fingers and eat them with your hands."

Next came the tossed salad with slices of boiled eggs, tomatoes, onions, bits of radish, crisp toasted garlic-bread fragments and sprinkled with pine nuts and crisp bacon along with a hot loaf of rye bread and tubs of butter.

"This is almost a meal in itself," Aron remarked.

"In fact," Larry said, "You can order a larger portion as a meal if you just want a salad on a hot day."

Appearing with a platter piled high with steaming beef ribs, the waitress put this between them and set two plates in front of them with their baked potatoes which were slathered with sour cream, chives and shredded cheese. Departing, she said, "Here you are gentleman, enjoy yourself. "I'll bring you a bucket for the bones. Do you want to take them with you?"

"Yes," Larry replied. "My family's dogs will enjoy them."

Both men attacked their baked potato to make room for the wonderful smelling ribs which were tantalizingly close to their nose. When they were done, Aron asked, "You first," and watched as Larry broke the top rack of ribs into two pieces and put one piece on his plate. He then pulled a rib off and grasping it between the fingers put a meaty side into his mouth and munched noisily on the succulent dish which tasted of beef, honey, oregano, catsup, mustard and citronella in a cascade of bursting flavors on his pallet. While still chewing he motioned for Aron to take the other half-side of ribs which were on the platter.

Arron assented but more carefully lifted a half rack and tentatively pulled on a rib until it broke free and then using both hands brought them to his mouth where he bit off a chuck. Chewing at first softly and then more aggressively he masticated the meat and discovered a mix of taste and flavors that he had never experienced before. By the time he had finished his first rib, Larry had demolished his and was breaking another half-side of ribs off the platter.

"Eat up if you want any," Larry said kiddingly. "Or I will finish this entire platter by myself."

"These are really good," Aron replied. "I have had ribs before, but never any that tasted like this."

After finishing the rib he was working on Larry explained, "This is a mix of Southern U.S., Mexican, Native American and German flavorings that is unique to this part of Texas. Everywhere you go it will taste a little different as each pitmaster makes his own sauce. The kind of meat, the age of the animal and what it was feeding on make a big difference too."

After they had demolished the platter of ribs and Aron retired to the bathroom to wash himself up. When he returned, he found at his place a slice of hot rhubarb-strawberry pie with a scoop of vanilla ice cream on top.

"What is this?" Aron asked.

"That is rhubarb which is available here until late fall, and I think makes a fine contrast to end a meal of bar-b-que ribs," Larry replied.

"You're right," Aron agreed. "This has just the right mix of sweet and sour to cap things off. I should warn you that after flying all day and eating this huge meal, I am going to pass out on you in the truck. I'm sorry that I am not going to be much company, but I have got to get a couple of hours sleep."

"You go right ahead," Larry agreed. "We are only half-way back to the ranch and the folks will want to meet you when you arrive."

Aron did as he threatened and was jolted awake when the truck rumbled and bounced as it crossed a cattle guard which led off the paved road into the startlingly different cactus covered rocky vistas. There were mesquite and cedar trees and a few scattered cottonwoods down in the dry creek beds they crossed, but most of what he saw was rocky limestone outcrops, more varieties of cactus than he had ever seen and scattered clumps of dry grass.

"It is different down here," Aron observed. "It is like we are in an entirely different world."

"'We are," Larry agreed. "The lush green grasses that you saw around Alpine and Marfa on those meadows are long gone, as are most of the trees except in the highest elevations which catch a little more water and snow. The animals are different too. We now have desert mule deer, javalena, a smaller whitetail, different turkeys, desert quail, ducks and geese on what little water there is, mountain lion, bobcats, coyotes and even an occasional jaguar. See, there is a roadrunner ahead of us."

"So that is what they look like," Aron said, "I had seen them in the cartoons, but that is the first one I have ever seen alive.

"How far are we from the ranch?" Aron asked. "I really need to take a shit after all that food."

"We are probably a half-hour away," Larry stated. "When we get to a level place you can find a bush. But before you drop your pants look around carefully. Everything down here wants to eat you. There is some toilet paper in the glove box, and maybe a package of baby wipes."

Larry stopped the truck and from behind the back seat and drew out a single-barreled 12-gauge shotgun. He opened the gun and stuffed in a load of no. 6 shot and walked away from the road ahead of Aron. Carefully walking between the ocotillo and prickly pear cactus he motioned to a patch of relatively open ground and pointed. He then turned and walked back towards the truck."

He was soon joined by Aron, "Thank you Larry, I feel much better now."

"Look at that bush ahead of us," Larry said. "Do you see anything?"

"No. Just a bush," Aron replied.

"Look for a black round dot towards the base of it," Larry suggested.

"No. I still don't see anything," Aron said.

Larry cocked and raised the shotgun. When he fired a jack rabbit that measured nearly a yard long from head to tail jumped out from under the bush and died instantly as its head had absorbed most of the shot.

"That will be part of breakfast tomorrow," Larry assured his astonished guest. "I am going to show you how to make a bag out of the skin. My dad always insisted that whatever we shoot, we eat or use in some beneficial way. I'll clean it as soon as we get to the ranch."

The first indications that they were approaching the ranch were that the fences and cattle guards were getting more frequent. Then on a relatively flat ridge an airstrip had been bulldozed out and a hanger built to house a workshop and a small aircraft.

"Dad flys," Larry said. "He bought and maintains this plain so that he can fly people out to Marfa when they need to go in an emergency and to help him look after the ranch. Periodically he will fly the fence line to make sure the cows have not broken out somewhere. Mostly that's no problem. They know where their food and water is and tend to hang around those areas, but we have to move them to keep them from overgrazing the same ground. That's why some of these fences intersect at tanks or water holes."

The rambling ranch house was a two-story adobe structure with Victorian add-ons such as a three-quarter wrap-around porch and a round turret room on one corner. The later additions were ordered from Sears and Roebuck and shipped cross country by train, hauled by the wagon load to here and erected as a kit. Otherwise, almost all of the other ranch buildings are made of local stone and used as little expensive dressed wood as possible."

Beside the range was the stable and horse coral. The horses were mostly standing in the shade of the barn, but looked up to watch the pick-up drive into the yard. When Larry got out of the truck, he was greeted by two dogs who enthusiastically came with wagging tails and sat expectantly to be greeted and possibly receive their doggy treats.

"Take nice," Larry instructed as he handed a pair of the rib bones to two Australian blue heelers and a yellow lab. "This is Hector and Achilles our stock dogs and Demeter our general-purpose hunting dog. When they get through with their bones, I will introduce you more thoroughly."

Walking over to the coral Larry was approached by a roan-colored mare named Lightning who stuck her nose over the top of the fence. As Larry ran his hand down the white-flecked snout he asked, "Has Louise been riding you enough?" he asked.

Aron would later report that he thought that Ligthtning understood every word, and when Larry asked the question shook his head as if to signal no.

While he hoped his meeting Larry's family would not be quite as traumatic as is encounter with his Georgia in-laws, he had mixed feelings as he and Larry approached the front porch and prepared to enter the house.

Chapter 2

Ranch Life

As Larry Jr. had flown down the week before, he had a chance to check on his college classmate, Albert, who was serving as an in-house rehabilitation specialist for Larry's father and mother. Instead of having to urge the older couple into doing set exercises, he was having more trouble restraining them as they slowly regained their muscle strength and balance.

Albert had been warned that Larry Sr. would be difficult to manage, and he had found that warning to be appropriate. He wanted to get out of bed and go riding, driving or flying before he had the leg and arm strength to control a horse or operate controls. This had included a scary trip out on the old ranch jeep that had somehow made it through World War II and had been more-or-less running around the ranch ever since.

With no power-assisted anything, the driving lessons had started out on straight sections of the ranch roads and progressively increased to rougher, longer trips as the old man slowly regained muscle function. During the early stages Larry Sr. would drive them out somewhere until he was tired, and then Albert would drive them back. Getting used to driving a stick shift in rough country in a vehicle whose shocks were blown to the point where Blair Adams, the ranch mechanic, had to replace them, was a new experience for Albert among many, many others. Albert was glad that Larry Jr. had come home so his son could go flying with with his dad. Something Larry Sr. had wanted to do almost as soon as Albert had arrived at the ranch.

When Larry Sr. said, "anyone can fly these Cessna 150s," the thought running through Albert's head was that "anyone can die in these Cessna 150s." He was not encouraged that he had to help the old man get out of the car, into a walker and help hoist him into the pilot's seat and then climb inside with him.

"Albert," Larry Sr. instructed over the headsets, "once we take off and level out, let your feet and hands gently rest on the peddles and stick. I want you to feel how the aircraft responds. The tricky part is not controlling the plane once it is in the air, it is in the take-off and landing without stalling it out or flipping it over."

Despite Albert's initial misgivings, he found himself flying the aircraft and saw that he was able to keep it at a given elevation and heading.

"See, nothing to it," Larry Sr. stated. "If we have to go somewhere you can do most of the flying, and I will handle the radio stuff along with the takeoffs and landings. Only one of us can be trying to control the aircraft at the time. Otherwise, we die. Understood?"

Albert thought that he had a more than adequate understanding of the "flying and dying" part of this airplane business, and assented.

They had just returned from their initial flying experience when he noticed Larry Jr., and a man he assumed to be Aron, standing by the horse corral, and walked over to greet them.

"Aron, I'm Albert," he said as he extended his hand to the taller of the two as he spoke to Larry Jr. "I just got back from flying with your dad, and I am glad that you're here to go with him next time. He did fine so far as I could tell, or at least we got back alive, and we didn't wreck the Cessna."

"That's Dad," he responded. "He waited until I was gone so he could drag you up there. He knew that I would not let him fly, if I had known anything about it."

"I suspected as much," Albert replied. "They are waiting for you inside and are anxious to meet Aron.

"Aron it is good to finally meet you too," Albert added. "We have indirectly crossed paths, as your wife was at Beth and Larry's wedding."

"Knowing what Sam and I were about to go through with having a baby, not being really a member of the family yet and the severity of the COVID-19 epidemic, I passed on that one," Aron explained. "As it turned out we both had mild COVID cases anyway and Michael turned out fine."

Even if Aron had no idea where he was before, looking at the ranch yard and house would have indicated that he was deep in the heart of Texas. A fence of ocotillo cactus supported by cedar post and single strand wires surrounded the yard. At the front was an arched gateway made of deer antlers with a silver tin star cut out of sheet metal encapsulated in the middle of the arch.

"Admiring the massive structure," Aron asked. "Did all these horns come from here?"

"These are antlers, because they are shed every year," Larry Jr. explained. "Horns, like cows and sheep have, grow throughout the animals' lives. All of these came from either my dad's ranch here or from mom's place next door. Dad, as much as possible used local wood and antlers to make furniture in the house. He felt like hanging another man's horns was like riding another man's saddle. It just wasn't right."

Twin doors with glass panels in the front fronted the house. The scraped and stained heart pine wooden moldings around the interior doors and nearly floor-to-ceiling windows were in the Eastlake style made famous by the resort in New York State. The polished wooden floors were covered with oriental carpets on top of which a modern rubber Treadway had been installed to provide firmer footing for the older couple.

When Aron walked in he found Larry Sr. seated in a rocking chair covered with steer hide while his wife, Helga, sat beside him in a more comfortable glider which was made of wood and antlers with upholstered pillows stuffed with goose down.

"It's good to finally meet you, young man," Larry Sr. said as he rose and extended his hand. "Helga and I have heard so much about you from Larry Jr. and Beth. We hope to give you a real West Texas experience while you are here. We won't try to make a cowboy out of you, but at least show you a few things."

"Larry tried to tell me a little on the trip down, but I am afraid that after we ate in Van Horn, he lost me," Aron reported. "I was worn out from getting up early to make it to El Paso at a reasonable hour and then with a stomach full of the best beef ribs that I ever ate I could not stay awake. I didn't wake up until we crossed the cattle guard."

"Beth and the new baby must be doing fine for you to leave for a couple of weeks?" Helga conjectured.

"He was a month premature, but he was a big baby at birth," Aron informed. "He is at the pre-crawling stage and is going to be moving around like a little human tank, according to my mother. All of the blood work came back negative, so his children won't have the problem that Sam and I did."

"Beth and I are all right so far as the chromosome stuff goes too," Larry said. "It's just that we don't have time for children right now. We plan on that in a few years when I can support us."

"Don't wait too long," Larry Sr. rebuked. "Helga and I are not going to be around forever, you know. It was close with me for a while, but thanks to Albert we are getting along better than we were and hope to be back up to full 'fit and fiddle' in a few months. We just got back from a flight."

"I know, Albert told me," Larry informed.

"With this much going on and with smugglers, drug mules and emigrants trying to get in, flying is the only way to find out what is going on the ranch in time to do anything," Larry Sr. explained. "It can take days for the Border Patrol or Sheriff to respond."

"Supper is going to be ready soon," Helga said. "I know you had a good meal, but the rest of us are hungry."

"Aron, one thing that you will find out about us down here is that we eat," Larry Sr. explained. "Marriages, births, death, graduations, holidays, family reunions, roundups, new houses, new jobs or anything that involves living is celebrated with food. Take as much or as little as you want, but you must take something so as not to insult your host.

We serve family style. If something is super-hot, Larry Jr. or Albert will warn you about it. We serve iced tea all year, and you may put sugar or honey in it if you like."

Once everyone was settled around the table, Larry Sr. said grace, "Lord. Bless this food that this land has produced for the nourishment of our bodies and for the fostering of all good works."

First to be passed around was a salad with lettuce from the Rio Grande valley and watercress from a nearby spring accented with boiled duck eggs and pecans. Being somewhat forewarned by wagon-bed table which threatened to sag with food, he put a tablespoon of the greens in his salad bowl. This was followed by a German style potato salad, and he added a spoonful of it to the bowl.

After this was consumed and the bowls taken away, a steaming cast iron platter of fajitas appeared along with bowls containing various sauces.

"The red sauce is mild, the green one warm and the vinegary one with the tiny peppers is very hot," Larry Jr. advised.

Aron tasted the meat, onions and pepper mix before adding a little of the red sauce to his meal.

"This is deer meat, but it could be anything," Larry Jr. said in response to Aron's questioning look. "That jack rabbit that I shot will be part of tomorrow's breakfast. The cook has already cleaned it, and I have the hide turned and salted prior to tanning it. We'll use it as part of our outfit."

Following the fajitas, there were steaming fresh green corn tamales which were so good he ate two of them, even though he was nearly stuffed. Cornbread, butter beans, relishes, sauces and finally coconut cake and coffee finished the meal.

"I knew there were palms in California, but I did not know that they were here?" Aron questioned.

"We can do a few dates and coconuts if the winters aren't too bad, but I think that the cook probably cheated a little bit here," Larry Jr. replied. "Dad, the liquor store in Van Horn did have some George Dickel Rye, and I picked you up a half-gallon bottle. I'll bring it in, and we can talk about what Aron and I can do next."

Having retrieved some old-fashion glasses and once again retired to the parlor, plans were plotted for the night and the next day.

"I'd like to take Aron and Albert out to the shooting range in the morning, and then go to the line cabin where we have the blacksmith's forge and stay there before we get started on our event with Reverent Pederson at The Solitario," Larry began. "There we can make our spears and maybe some bolos and other gear."

"So far as I know," Larry Sr. said, "there is no one up there. You can take your horse up, and Albert can drive your gear and food up and then go back and get you."

"Aron, have you been on a horse recently?" Larry Jr. asked.

"No. I don't really know anything about them," Aron said.

"I don't want to make you more miserable than you are going to be by putting you on a horse until we have more time," Larry Jr. explained. "We are going to work on toughening your feet and legs, and you don't need horse-sore muscles to go along with them."

Aron had watched and observed how the men slowly sipped at their whisky before trying some himself. When he did he remarked, "This is really good. It is strong, but really has a fine, smooth finish and taste."

"Life is too short to drink bad whisky, despite what you have seen in the movies," Larry Sr. replied. "I have been offered some of the world's best, and the Scotches are too oily and smoky for me. This is far better as they chill it to take the fat off it, which means that you have fewer ill effects from drinking it."

"So far as sleeping arrangements go," Helga informed, "Aron there is only one bed in the house that is long enough for you. Years ago, we had a traveling preacher, Parson Emery, who would come through. He was a tall fellow and it was made for him. We cleaned that room up for you. The bath is down the hall and Larry Jr. can show you."

"I am sorry, but I am losing it again folks," Aron responded. "Anything that is not moving, biting or burning would do good right now."

Larry Jr. thought, "You are more right than you know, brother. Soon enough you are going to get all of those."

Before dawn the next morning, breakfast saw Larry Jr., Aron and Albert at the table looking at a pile of corn pancakes with sides of cut fruit, syrup and tortillas filled with the cooked jackrabbit smothered in a sweet-sour sauce topped with a dash of cream fresh.

"Your rabbit turned out to be tinder and good," Aron observed.

"The cook marinated it and then slow cooked it nearly all night on the back of the stove, and then made up the sauce for it," Larry Jr. reported. "The younger ones are not nearly so tough and are reasonable eating, although not as tasty as the smaller cottontails. Before we are done you will probably have the chance to try both. You and Albert finish up. I want to saddle up and meet you at the range. Albert, the guns and ammo that I want to take are in those two cases in the hall by the door."

On the ride out to the range Aron asked Albert, "How are you making it through all of this?"

"First at Sam and Beth's father's house in Georgia and now here, I have had the best experiences of my life. I am an orphan and was raised in institutions of one sort or another," Albert explained. "This is the first time I have been treated like a person. I feel close to these people – like they were really my family. I don't know if I have adopted them or they have adopted me, but something like that has happened."

"I had encounters too, especially with Elizabeth and Margaret," Aron remembered.

"Me too," Albert said with a broad smile involuntarily flashing across his face. "The range is just over the hill I think."

The shooting range had been made from scraped out dozer scars and earth berms at 25, 50, 100 and 200 yards. A shooting porch shaded three benches and slats in front of and behind provided another degree of shade. Lightning was tied up under the shade of some cottonwoods near the creek where a trickle of water was running down from a spring further up the hill.

Larry Jr. opened up the two gun cases and revealed a variety of guns and boxes of ammo. Knowing that neither of his guests knew hardly anything about firearms he started with the simplest of them, the single-barreled shotgun that he had taken the jackrabbit with.

"We keep inexpensive shotguns in all of ranch trucks," Larry began. "With slugs they are used to put down sick or wounded animals, with buckshot they can take deer, coyotes and mountain lions and with birdshot quail and ducks. With buckshot these are devastating on a person at close range. If you have shells in a pocket or in your hand it can be reloaded very rapidly. If you are going to hit what you are shooting at; however, they must be aimed. A shot fired in the general direction of an opponent will likely not hit him or do serious damage. This is a gun to use at 20-yards or closer.

"Next, I have you a 10mm. Ruger Super Redhawk. This is a revolver, but uses these clips to hold the ammo, although you can shoot the gun without them. I brought this one because it is powerful enough to do some good on anything you might have to shoot, but does not have the recoil of the Remington .44 Magnum. This is an accurate low-recoiling learner gun which has a scoped sight. You put the crosshairs on the target and slowly squeeze the trigger and you will hit it.

"Third up is an AR-15 style of gun in .308. This will kill anything that walks in this country, will penetrate vehicles, gives rapid-fire capability and if we ever have to defend the house against anyone this is the style of guns that we have. It can be used with or without a scope, and I would like for you to practice with both.

"I am going down to set up some targets. Do not touch any of the guns while I am down range."

"Have you ever shot anything?" Aron asked Albert.

"Not even a BB-gun," Albert replied. "Guns or any sort, even rubber-band ones were the last thing that they said they wanted to see around the orphanage."

"I, at least, had those and a water-gun once, but that was as close as I came to shooting a gun," Aron concurred. "Even when we went to the fair and there was a shooting gallery, I was never allowed to touch one."

With earplugs and shooting glasses, the pair were taken through the operational characteristics of all of the guns. They concluded with both Albert and Aron hitting man-sized targets in vital areas at 150 yards with both the pistol and rifle.

"You did well for a first-time out at the range," Larry Jr. complemented as he unloaded the guns and returned them to their cases. "Albert you can take Aron to the cabin and stay in the truck until I arrive. If anyone is there, turn around and come back. Lightning and I will be a few minutes behind you."

When everyone arrived at the cabin, they found it to be empty. This was a stone-walled single room structure with a fireplace at one end and a tin roof. When Larry Jr. opened the door he was pleased to find that its only recent occupants had apparently been some rats and bats.

"Unload the stuff on the porch and put the coal by the forge," Larry Jr. instructed. "I want to start on making those spears tonight while the weather is cool."

Larry Jr. busied himself sweeping down the metal springs on the bunk bed and sponging off the rubber mats that were on top of them. Then he got the camp broom and after opening the windows proceeded to sweep the walls and floors which raised enough white dust to choke him to the degree that he had to step outside.

"That limestone flag floor is solid, but it is hard to sweep," Larry Jr. told the astonished Aron. "It beats having a wood or dirt floor though – fewer critters you see."

While Aron got some pots and plates washed in the creek and the small box stove washed down and loaded with fire wood, Larry Jr. started stoking up the forge. He needed to get the steel he had brought with him up to a white heat to make the two spear heads that he would affix to the ash shovel-handle shafts.

These tool handles were big enough and straight enough to use as spears, but the problem was attaching the metal spear point to the shaft using the scrap steel that he had. These were drive shafts from old lawn tractors that were half-inch in diameter, but had gears and connections attached to them. The tool handles were tapered and split at their ends and with the rough equipment that he had he doubted that he could produce a flat enough and wide enough piece of steel to fit into the shafts. If he tried to wedge the steel into the wood, he might split the shaft and weaken the entire spear.

Knowing it would take only a few minutes to heat the canned beans and spam they had brought to the camp, Aron joined Larry Jr. at the forge.

"These bars have enough material to make two spears. The first thing I am going to do with it is to get it hot enough to remove these excess metal parts, so I have a piece of straight round stock. Then I will heat it to a white heat, forge the blade and then flatten the back. The point will be about a foot long with a leaf-shaped tip. Its shaft will be circular down to where there is already a shoulder. I will square off the shoulder, drill a hole into the shaft with a battery-powered hand drill and burn the shaft into the handle and then cross-pin it into the shaft. This will be something on the order of an African throwing or a Roman spear. Then I will grind it on the wheel until we can get a keen edge. I will also make us some small knives.

Digging out cowhide gauntlets that reached to his elbows and putting on safety glasses, Larry Jr. started to work the metal sending showers of sparks flying on the dirt floor of the forge. "Watch that none of those catch anything," Larry Jr. instructed.

After pounding the bar more-nearly square he proceeded to shape the twin edges of the spear point. When he was satisfied that the edge was as thin as he could work it, he started pounding on the back part of the bar to make the 4-inch tang that would be fit into the shaft. "I have to be careful, or I could break the spear blade off at the head. The stronger socket fitting that was used on both the African and Roman spears is difficult to forge, and I need a cone-shaped die to work the metal. The tapered socket is stronger, but I don't have the tools for it. The results won't be beautiful, but these points will be more effective than stone and make a better-balanced throwing spear."

By the time the blades were forged it was after dark and the pair pulled up chairs to the table and consumed their beans and sausages.

"These taste a little different than I am accustomed to," Aron remarked.

"This sausage is chorizo and uses a different spice mix than the typical sage, salt and pepper used in American-style sausages and wieners," Larry Jr. explained.

By the time the pair had completed their stay, the spears and knives had been completed and a coyote was killed and skinned to make sheathes for their knives and a carry bag.

"You remember the Ice Man that they found frozen in the Alps?" Larry Jr. asked. "We are going to be carrying the same tools - our hunting and butchering items, fire-starting punk and flints, dried foods and some gourds of water. Hopefully, we will not be hunted down and shot like he was.

Chapter 3

Survival

At the Full Gospel Bible Camp and Nudist Colony outside of the old mercury mining town of Terlingua, Reverent Pederson oriented the two men about their coming experience. "The volcanic crater is 14-miles across and you will be there for seven days. Here are two maps showing the five cache points which are piles of stones painted in different colors. Bring a piece of each colored stones back with you to complete the course. At each of these caches are four bottles of water. There is also water at each of the three stock tanks but you should boil that water before you drink it unless you take it right out of the pipe from the windmill.

"There are ruined buildings at the old mine in the center of the crater. You can use anything that you find there and can salvage. That is also the best water source, although it has lead and other heavy metals in it. The better drinking water is from the windmill near the road where we will drop you off.

"There are no cows in there, but a stray could always wander in. There is the usual Texas wildlife which you can take and eat. One or more mountain lions may be there too. You will also very likely hear and see coyotes. I don't know if we have El Tigre, our jaguar, around at the moment; but who knows."

"There are rattlesnakes that have evolved to the point where their colors match the rocks, so they can sometimes be hard to spot. There are also scorpions of various varieties of which the smallest is the most poisonous. You will also likely see some of the large tarantulas, which are terrifying to some people, but are harmless.

"On the plant side almost everything out there will prick or stick you. The cactus, particularly the ocotillo will stick through your blanket. This is peyote gathering season, and those will give strong hallucinogenic responses. The only fatality that we have had was from a young woman who thought she knew everything who apparently ate some. We found her bones 30 miles away. It is assumed that whatever visions that the cactus brought were so terrible that they drove her to her death.

"Be particularly careful with fire. As dry as it is a wildfire can range through this country and get you before we could ever pull you out.

"There is an emergency phone at the mine, if anything happens and you need to get out early. The telephone number that you need to call is on the map. If for some reason that does not work call 911 and ask them to relay a message to us through the sheriff's department. Otherwise, we will see you at the mine in a week. Do you have any questions?"

Having talked with Larry about nearly nothing else but this experience for the last several days, Aron mentally ran through their conversations and could not think of anything. "Nothing from me," Aron replied. "I think Larry and me have just about covered everything."

"Maybe not," Reverent Billy retorted as he threw Aron a bottle of very high blockage sunscreen. "At the colony we find that visitors come with tans, but forget that these rocks reflect sun rays on their undersides and can be very badly blistered between their legs and on their bottoms. Slather up well before you go."

"Thank you. We'll do that. I don't need toasted balls too," Larry Jr. quipped.

With their tennis shoes, blankets, now serapes, tightly gathered around them with their knife belts, the bags thrown over their shoulders they began the descent from the road entrance to The Solitario to the mine buildings below.

"Larry, is this what you thought it might be?" Aron asked.

"Just about," Larry Jr. answered. "According to the map one of our caches is not far off the road to the east. We can visit that one and hunt along the way. We need to get something for our supper tonight. We'll go slowly down the road, hunting as we go; and then go to the cache, back to the road and then down to the mine.

As they had been put out early in the morning the heat began to build up as the sun climbed higher above the rim of the caldera. "The animals here mostly move during the cooler parts of the day and night then they water and bed down in whatever shade that they can find," Larry Jr. explained.

The rhythmic metallic noises of the clanking windmill could be heard above the wind. They had spotted its flashing blades almost as soon as they got out of the truck. As they approached, they spotted a buck mule deer coming to water and the pair watched it as it watered and walked away.

"Game will often use the same approach and exit paths to water. In this dry noisy country you can't often stalk game. What we will do is to set up blinds on either side of the trails and see if we can ambush something that walks within throwing range of our spears. Before we even try to take a deer, we have to be ready to salt, smoke and dry it or otherwise the flys, hornets and bees will get it before we do to say nothing of the coyotes and buzzards. We'll feed them too, but not with the meat we hope to eat."

The pair quickly assembled two brush blinds so that they overlooked the trail and gave them a down-angle throw within 10 yards of the trail. This positioning prompted Aron to ask, "This is so close. Won't they smell us?"

"We should let these blinds stay for a couple of days before we hunt them," Larry Jr. responded. "Which one we use depends on the wind. It is very variable in this caldera. As the air warms it starts to move from the bottom of the caldera to the rim, whereas in the morning the dominant flow is from the rim down. If we get hard winds blowing over the top, almost anything could happen so far as wind-direction goes."

After they had finished making their blinds, they started out to the first cache. "When trying to navigate, the best thing is to pick out a distinctive point on the rim and head for it," Larry Jr. advised. "You can break brush, pile rocks and look for your footprints, but in this dry, brushy country you may be walking on rock and not leave any prints at all. If you keep going downhill sooner or later you will wind up at the mine. In this quadrant you will intersect the road or get to a place that is high enough to spot the road or windmill. Your gut sense of location or distance will often fool you."

Aron noticed that while he was walking Larry Jr. had been picking the purple fruit from some of the larger prickly pear and putting it in a bag. Noticing his curiosity Larry explained. "The juice from the cactus fruit has sugars and also vitamin C. I crush and strain it and then we will have a little something sweet to drink and something besides meat to eat. Once you remove the spines, the cactus pads themselves can be boiled and eaten. They taste somewhat bland, something like string beans. There is food all around you, but you must be careful about it because some of these plants are deadly."

Soon they found a yellow painted pile of stones and each of them removed a pebble to put in their bags and recovered the water bottles that were also in the pile which also contained a bar of soap.

"I see that they also have put out a few extra goodies for us," Aron observed.

"That gives us some added incentives to hit all of their caches," Larry Jr. remarked. "Let's go down to the mine and set up housekeeping and see what we can salvage down there. Then we don't have to carry everything around with us."

The headframe of the mine still stood along with a scattering of other stone walled and wooden buildings. The largest was the two-story mill house, which had contained the machinery to crush the ore and recover the galena. Its roof was gone, and the second story had been removed, but the walls and concrete floor were still present and had been cleaned out. There was also an office which still retained its tin roof and another building with a stout chimney which was likely the cookhouse. Remains could be seen of a wooden barracks, the foreman's house and of smaller sheds but these were mostly piles of weathered planks.

"We can use some of the broken-up wood for fires," Larry said, "but those weathered planks have value for making rustic paneling, picture frames and cabinets."

Looking in the cookhouse they found that this was where people had obviously stayed. An open fireplace had been built into the chimney and on the hearth was a variety of pots and a swinging hanger with a hook to suspend a pot over the fire. In a cupboard they found salt, rags, brass pads for cleaning pots and a small first aid kit that only contained a few Band-Aids and iodine. Otherwise, the room was empty, and they spread their bedrolls out on the concrete floor.

"There were some kerosene lamps, but I suppose they all got took by someone who needed them worse than us," Larry Jr. stated.

"I guess we are going to rise when it is light and sleep when its dark," Aron observed.

"Yep. That's about it." Larry agreed. "We have about half-a-moon tonight so unless it clouds up there will be some light. Let's see if we can find anything useful in the other buildings."

Aron found that he had a strange, foreboding feeling walking around an area where scores of people had once been intensely working amid the roaring sounds of belts dumping ore and massive stamps crushing it and rotary mills grinding it, but now only scant traces remained. In protected corners rats and other rodents had nested and spiders had spun their webs which were now mostly covered with brown dust. In a pile of rocks beside the shaft Larry Jr. picked up one and showed Aron a clump of sparkly white galena which had cleaved into cubes.

"This is what they were looking for," Larry said. "This is galena, and it sometimes contains significant amounts of silver. I don't guess that they never found enough of each to actually work the mine. They spent much more money building the mine and mill than they ever took out of it. Mark Twain, who speculated in silver mining in Nevada, once said, 'A mine is a hole in the ground with a liar at the top.'"

Seeing birds hovering over the mine shaft and diving in Aron asked, "What are they doing?"

"Those are mostly quail," Larry Jr. answered. "They dive into the shaft, sit on planks floating on the water, drink and fly out. If we make a net, we can maybe catch some. There was a working windmill here too, but because of the lead contamination the rancher moved it to the other side of the crater where it is pumping water to the tank over there where we can get our water." As Aron spoke, he pointed to an open welded steel tank offset a bit from the mine buildings and above a small green patch where water flowing from the top of the tank kept that small set of Texas plants alive.

Aron felt a tinge of hunger in his stomach which reminded him that it was past midday, and they had not eaten anything since breakfast. "I'm getting hungry," Aron remarked.

"It looks like we had better be about finding us something to eat," Larry Jr. agreed. "Let's check over by the tank here and see what's been coming in."

When they looked at the green patch below the steel tank, they discovered that the ground had been heavily rooted up.

"It looks like there are some javalina," Larry Jr. related. "They are probably lying up somewhere fairly close. They are vocal and noisy when they move. Oftentimes I am able to hear them before I see them. Usually, the smaller ones will come out first, and then the larger ones. Look for some wire or scrap cable, and I will make a snare and see if we can catch one."

Finding a rusty six-foot length of rusty cable, they untwisted two strands and using rocks beat them out until they were once again reasonably straight wires. "The snare is fairly simple," Larry Jr. explained. "You set the loop so that the animal's head or foot sticks in it and then when it moves forward it tightens. The other end is fastened to something large enough to hold it. As it attempts to move the snare will tighten around the animal. With what we are using we are going to have to be on it in a hurry, because these twists that I am putting in the wire may not hold an animal for long. We will kill it with our spears."

Larry explained that the spears that he built could be used either as a club as a sort of extended knife for slashing, stabbing or thrown. The best approach was to use it so that the point was always pointed away from you and not attempt to club anything with the butt unless the spear was nearly vertical. "Otherwise, if you are swing a spear wildly you could stab yourself or me in the excitement of trying to kill something. Once you throw your spear you are defenseless except for your belt knife which is too short to really be a fighting knife."

It was not too long after that squeals were heard from the vicinity of the water tank. The pair grabbed their spears and rushed over to find a javalina tugging against the snare. Aron hesitated, but Larry Jr. urged him, "Go and kill it with a thrust behind the shoulder."

For the first time in his life Aron was going to kill something and do it in a close and personal fashion. This was not to be an antiseptic shot with a rifle at 100s of yards, but at a matter of feet. The javalina saw him and faced him. Larry came up beside him. "We need to get him from behind the shoulder. You work around to the side until you can spear him from the back."

The pig would have none of it. With two adversaries facing him he turned and broke sideways. Larry reacted swiftly and threw his spear hitting the animal in the neck which slowed its forward progress enough so that Aron could plant his spear point between the shoulder and ribs and drive the point though the animal. Bleeding and gasping from two hits the javalina quickly bled out and died.

Aron was hanging onto the spear whose point having passed through the animal was planted into the ground. He felt like he might be going to faint. Larry Jr. came over and supported him while Aron tried to mentally process his emotions.

"I didn't know. I didn't know it was going to be like this," Aron began. "I don't know how I am supposed to feel. I'm horrified at what I have done, sad to have killed the animal, glad that we are going to have something to eat and proud of myself for doing something that I did not know that I could ever do. It wasn't fun. I don't know what it was?"

"Expressed in terms that Native Americans might have used," Larry Jr. began, "you are on the pathway towards becoming a Brave - one who can survive and support his family. This is one step, this first kill, towards meeting that objective. You empathize with the animals you take, you honor them by using as much of them as possible as you feed and protect your family. This is not sport in the Western use of the word, it is gathering food. You should take pride in your accomplishment in facing this aspect of death and life." When he concluded he dipped a finger in the animal's blood and stroked him on both cheeks. Aron was now a blooded hunter.

They took the animal away from the water tank and Larry Jr. hung it from a limb, gutted it and skinned it. The smallish animal weighed about 30 pounds which would yield relatively tender meat. "The critters will spot and clean up that gut pile tonight." Larry Jr. informed. "In the meantime, we will cut up some cactus and make a stew from some of the meat and roast the hams, shoulders and ribs in the fireplace."

In an attempt to make the meat as tasty as possible Larry Jr. crushed up the cactus fruit and added that to the pot along with strips of prickly pear from which he had cut out the spines. "As a meat, javalina is not very tasty, but it will keep us alive," Larry Jr. commented. "I am putting some salt on the hide tonight and tomorrow we will scrape it, oil it, smoke it and cut it into thin rawhide strips to make a net. The best leather driving gloves are made from javalina hides."

While Larry Jr. had been processing the animal, Aron was sent to cut the tallest grass that he could find to make two sleeping mats for the night. As they spoke they wove the grass stalks together. Beneath the mat they placed a layer of small juniper limb tips to provide an aromatic mattress. Aron did most of the weaving and found himself surprised at the amount of grass it took to make a single mat. He found the repetitive work enjoyable as he interlaced and tied off the grass. "I never gave much thought to it," Aron remarked, "but I never considered having something soft to sleep on as anything out of the ordinary. Now I see how much work actually goes into having something to lay down on."

When even the light given off by the fireplace became too little to do anything useful, they laid on their improvised bedding wearing their sarapes. Despite their primitive sleeping conditions, exhaustion won out and they both dropped off to sleep.

Sometimes during the dark of the night Aron felt something run across his chest. He woke up with a start, and exclaimed, "What's that!"

"What's what?" Larry Jr. replied.

"Something just ran across my chest," Aron answered.

"How big was it?" Larry Jr. inquired.

"Big enough that I could feel it." Aron responded.

"Probably a pack rat," Larry Jr. informed. "If it were a large spider you likely would not feel it unless it was on bare skin."

Neither of the two options was very appealing to Aron who picked up a piece of wood by the fireplace and put it beside his bed before attempting to go back to sleep.

Aron heard something moving and looked to see that the cabin was again somewhat lit by the fire that Larry Jr. had started, and daylight was coming through the window.

"Go ahead and open the door," Larry Jr. suggested. "That will give us a little more light in here."

When Aron returned from outside, he left the door open, and some relatively cool air entered the building.

"How do you feel?" Larry Jr. asked.

"Stiff, tired and sore," Aron indicated by pointing to various parts of his body as he spoke.

"After we eat we can wash up, and I will get some aloe to put on our scrapes and pricks," Larry Jr. informed. "How did you and our pack rat make out last night."

"I think that he probably likes this bed better than I do," Aron replied. "I don't know if he came back or not. I finally just went to sleep."

"After we eat and clean up, we'll go out and see if we can find another cache or two while it is comparatively cool," Larry Jr. suggested. "We may find some other goodies that have been left for us.

"My question for you is do you want to continue?"

"Yes," Aron replied. "Everyone has gone to a lot of trouble to put this together, and I want to finish it."

"Good enough," Larry Jr. responded. "We'll go out and see what we can find."

Chapter 4

Capture

Thinking to locate one or two more of the caches during the relatively cool early morning hours, Larry Jr. and Aron busied themselves finishing what remained of last night's meal and prepared a slow-burning fire of green wood to continue smoking their remaining meat.

"You are right about this javalina," Larry noted. "It's sitting overnight has not improved it. It doesn't taste like any medicine that I ever had, but I would not want to take any medicine that tasted like it either."

"There is a scent gland above the tail that needs to be cut out as soon as the animal is killed," Larry Jr. informed. "Some animals' scent glands are used in the perfume trade, but not this one, I think. That smell is rank enough to cover almost anything."

Laying out the map, Larry Jr. could see that there were two caches more are less on the same arc near the rim of the caldera opposite the entrance road. Pointing to the two locations he said, "We'll try to hit these two today. I am going to pick a point on the rim to head for as we work up and follow game trails to get us there. The cows find a little more graze near the edges of the caldera because more moisture is trapped in the shaded areas. We'll hunt along the way and see if we can pick up a rabbit or something."

Even with tennis shoes they found that sharp-edged rock fragments could be felt through their shoes. "I can imagine what this would be like trying to do barefoot," Aron replied.

"Human feet do callus, and that helps," Larry Jr. responded, "but even the Indians made shoes and leggings to wear in this country. Attempting to walk barefooted would cut up your feet to say nothing of causing blisters."

They saw deer, but the noisy footing and variable winds prevented approaching them from close enough to use their spears. They also busted two jackrabbits. Although one stopped as soon as Aron started to wind up his bolos for a throw, the rabbit thought better of the idea and hopped deeper into the thorn bushes.

Empty handed, they reached a vantage point on the rim which allowed them to view to entire crater. Spreading out the map, Larry Jr. oriented it with the features and determined that they were likely about the correct elevation, but somewhat south of the cache.

"The people who set out these caches have to reach them too," Larry postulated. "I think that we will likely find them along the better-established trails."

Proceeding cautiously they located the next cash which was marked by green-colored rocks.

"This is almost like opening a Christmas present. What have we here," Aron said as dug out a tiny hatchet with a 11-inch handle and 2 ¼-inch blade and handed it to Larry Jr.

Removing the leather blade cover, Larry Jr. examined the blade and read, "Gransfors Bruks – Sweden," stamped on the handle. "This is the European take on a tomahawk. They are expensive, but are among the world's best tools of their type."

After drinking a little water and admiring their prize they continued traveling around the rim towards their next objective. As the rocks warmed the wind changed direction. Larry Jr. motioned for Aron to look below them and they spotted a jackrabbit grazing on a clump of grass. Aron got out his bolos, and started swinging it over his head for the throw. Hearing the unusual sound, the rabbit sat up on his haunches and presented nearly a two-foot-tall target for Aron's throw. The three balls and string rotated through the air like a Frisbee. When it hit the rabbit the balls and cord wrapped around it and it fell over kicking. Aron ran up and clubbed it with the butt of his spear.

"Excellent," Larry Jr. complemented. "That will give us some better tasting meat for tonight." As they picked up their rabbit a whitetail doe rose in front of them and ran with her tail giving them white-flag salute as she bounded away.

"Better a rabbit in hand, that a deer in the bush," Larry Jr. philosophized. "I was so concentrated on that rabbit that I did not see the deer bedded down behind it."

Soon something red attracted their attention and they approached the object. Larry Jr. examined it. "I think that I know about this one. This was a civilian aircraft that crashed in a fog. Someone, maybe the accident investigation team, helicoptered the wreckage out, and I suppose this piece fell off.

"We could hammer it out and use it like a wok if you want to carry it back. Once the paint has burned off, we can cook on the unpainted side. If we keep moisture in it, the metal will not get hot enough to melt.

"I think our cache may be above us and a little further along. Let's climb up and take a look."

About 15-minutes later Aron pointed out something metallic blue that contrasted against the greys, whites, tans and green in the landscape. This cash contained more water bottles and a small 6-power Leitz monocular.

"These will be useful to spot game," Larry Jr. commented. "I will say one thing about Reverent Billy, he is putting out quality goods."

"Who's that," Larry said as he saw a cloud of dust kicked up by two vehicles coming over the road. "They are going to the mine."

"I don't know," Larry Jr. answered as he focused the monocular on the objects. "They could be anybody. They are going to find out in a hurry that we are here. There is a pick-up and a tractor-trailer. These don't look like ranch trucks. They could be smugglers."

Larry Jr. watched as a group exited the pick-up and the back of the tractor-trailer.

"Are they looking for us?" Aron asked.

"I don't think so," Larry Jr. replied. "Whatever they are doing, we don't want to be found by them. We've got food and water and can keep an eye on them. Maybe they will leave tonight or tomorrow. There is a rock shelter over the rim called *la casa de Piedro Negro* after a black sheepherder who stayed there. We can spend the night and have a small fire without them seeing it. Tomorrow we will try to find out what's going on."

The explosion that had formed the caldera had uplifted and tilted a ledge of rock so that it overhung an open space about 10-feet long and seven-feet high. A dry-laid rock wall had been built between the shelter's roof and the ground to help block off the front. In function it was like living under half of an A-frame tent. Dirt had been hauled in to form a floor which was covered in animal dung, and a circular rock-walled area with charcoal at the bottom indicated where the fire pit had been.

Larry Jr. cut some juniper branches and using these they swept their floor as best they could. After this Aron gathered more branch ends to form their bedding while Larry Jr. picked up wood for their night's fire. The aluminum sheet was un-rolled and put under the junipers to add another degree of separation between them and the floor. As there was no door, this would be a cold night, although the fire would help warm the little structure.

After the new hatchet was used to split up the wood, the fire was started by Larry Jr. He struck sparks with a piece of milky quartz vein material and the steel blade of his spear. He had gathered some dry duff from rotted wood and blew up a glowing coal and then a flame by carefully adding a small piece of toilet paper that Aron contributed and then leaves and twigs. When a burn was achieved, he progressively added larger pieces of wood. The rabbit pieces were spitted on sticks, toasted like marshmallows and shared between them.

"Tough chewing," Aron remarked.

"Like you saw at the ranch, it can be boiled tender, but the more you cook it the dryer and harder it gets," Larry Jr. said as he worked his jaws as he masticated some of the long thin muscles from a foreleg. "This could really use some salt, but all we have is at the mine."

Through the night they slept back-to-back wrapped as tightly in their serapes as they could manage. Their growling stomachs attested to their hunger, but outside of water they had nothing to put in them. Larry plucked and cleaned some prickly pear cactus leaves and fruit and discovered some fiddle-head ferns growing near a small seep and added those to their breakfast concluding, "We need to eat what we can."

Peering over the rim at the mine below, they could see smoke rising from the office building's chimney. "What we have found in these caches has been really useful," Larry Jr. commented. "I want us to sneak around to the next one which is located about half-way down into the caldera and closer to the mine. Then we can conceal ourselves and watch until dark. They are busy now with their breakfast and hopefully would not notice us, although they certainly know we are here."

Moving rapidly at first and then more cautiously, they found a game trail that led towards the mine headed in the general direction of the cache. Hiding behind brush and small trees, they were somewhat less exposed than when they were silhouetted against the rocky rim, but they dared not speak above a whisper because they could begin to hear Spanish speaking voices emanating from the mine site.

Aron turned his palms out in a questioning motion to ask what they were saying, and Larry Jr. whispered back. "The leader is telling them to go out and find peyotes and collect them. This has been a dry year with scattered rain which is what the cactus likes. They don't look like much, but they bring serious money on the drug market because they have a higher mescaline content. We've got to find that cache and get out of here. They are going to be walking these trails just like we are."

Looking in haste Aron spotted a purple mound of rocks. When they unstacked the rocks they found a take-down recurve bow and string, but no arrows. Quickly gathering their prize, they returned the way they came, but stopped before they started to cross a series of open areas immediately below the rim.

"We can't go any further now," Larry Sr. observed. "We've got to wait until they are distracted."

Walking through the thicker brush and cactus they concealed themselves as best they could for two hours as they heard people moving below them. There were two shots, followed some five minutes later by a third.

"I think that they've shot a deer," Larry Jr. observed. "While they are fooling with it, we can make a break for it."

Stepping up their pace they quickly crossed the open ground. Once they breached the other side of the rim, they headed for their rock shelter. When they approached Larry Jr. stopped. "Someone could be here," he whispered. "I want to watch it before we go down."

Larry Jr.'s intuition was correct. Coming out of the shelter was a Mexican boy about 10-years-old with a sack and a short grubbing stick who was nervously, sitting, standing and pacing in front of the shelter. Satisfied that the boy was alone, Larry Jr. left his spear behind and approached. "Hello. I am not going to hurt you," Larry Jr. said in Spanish.

Whoever the boy might expect to see stepping out of the bushes it was not a filthy bearded Anglo apparently dressed only in a sarape. Although it was obvious that the boy's first instinct was to run, he stood his ground. Whatever terrible fate awaited him here, it was apparently better than what he had left behind.

In as sympathetic a voice as Larry Jr. attested, "Let us help you."

Jesus Sanchez, apparently relieved, sat on a nearby rock and involuntary tears started to flow. Larry Jr. approached, put a comforting hand on his shoulder and lifted his head and hugged him.

Seeing Aron approach with the two spears the boy's eyes opened wide in fear as this new, to his eyes, huge apparition approached who looked even more dangerous than the person holding him in his arms. If desert demons existed, this surely must be one.

Aron offered him a water bottle which he drank from and then resealed the cap. Be they friend or foe, the boy realized that he had no choice but to seek help from these strange men.

He said that his father, mother and sister were attempting to escape from Guatemala and had paid these smugglers to take them to the states. They were packed in the tractor-trailer with others and passed through customs because the rear quarter of the trailer had been packed with crates of avocados while they and the armed men had been in the front. Once across the border in the U.S. they had been forced to gather peyote and this was their second stop. They had been told that if they did not, they would be sold into the sex trade. His father had tried to defend them and been beaten up. The boy said that he had been whipped and told if he did not come back with at least 20 cacti he would be whipped again.

"Tomorrow night tell your mother to put peyote into the venison stew and feed that to everyone," Larry Jr. plotted. "When that takes effect, we will try to get into camp, free you, take one of the trucks and get you out. Go now, gather your peyote and get back to camp."

"How is this going to work?" Aron asked.

"Tomorrow evening after they start back to camp, we will make our move and find the last cache which must have arrows in it," Larry Jr. explained. "After the peyote takes effect, we'll rescue the family, disable the other vehicle and escape."

Sabastian was the first to return to camp after gathering his quota of peyote. He was thirsty and hungry and went into the office building where he found Senora Sanchez stirring a pot of venison vegetable stew.

"I'll have a bowl of that," Sabastian ordered.

"It's not ready yet," Senora Sanchez replied.

"It smells good to me," Sabastion affirmed. "I'll have some anyway."

"It's not done," Senora Sanchez protested. "It'll make you sick."

"It looks fine to me. Put some in a bowl and bring it here," Sabastian said as he took a slug from a bottle of hot beer.

Senora Sanchez brought him a bowl of the steaming stew, and Sabastian age hungrily. Soon he felt himself adrift, somewhere far away, beautiful, and peaceful. Wherever he was he wanted to stay. There were no demands for him to be anywhere or do anything – just existing was sufficient.

Jefe Juan Gonzales walked in and saw his henchman sitting wide-eyed staring into apparent nothingness, and demanded "What's happened to him?"

"He's drunk," Senora Sanchez replied.

"He's had peyote, hasn't he?" Gonzales demanded.

Gonzales picked up the bowl and looked at the residue and sniffed it. "You put some in the stew, didn't you? We'll have your husband try some and see how he likes it."

Walking outside he ordered Pablo Sanchez to be untied and brought in. Taking a spoon, he put more stew in the bowl and handed him the spoon and the half-drunk beer. "Eat," he ordered.

Pablo hesitantly drank a swallow of the beer and blew on the soup to cool it. Gonzales watched intently as his captive finished his bowl. Soon Pablo looked around, but his eyes seemed to not be fixed on anything in the room. He was apparently in some other place having some other experience than this room offered.

"Who put you up to this? Tell me. Tell me now!" Gonzales demanded. Taking the empty bowl he dipped it into the pot, ripped the back off Pablo's shirt and threatened to pour the boiling liquid on his bare skin.

"It was the Anglos," Senora Sanchez said. "They are coming tonight to get us."

"And this is how you repay our kindness?" Gonzales said as he motioned for his to men to hold Pablo down while he poured the boiling soup down his back. Pablo thrashed violently to throw the sticking meat and vegetables off his skin, and passed out before he was drug away.

"Shame to waste good stew. There is enough for you and your children too, if you don't help us catch them."

Seeing men apparently staggering around the buildings and Senora Sanchez walking out in the open and waving to them, Larry Jr. and Aron walked toward the mine buildings. When they approached the office, they were surrounded by men holding pistols and rifles.

"Drop your weapons," Gonzales shouted. "You are worth more to me alive that dead, but that doesn't mean we can't shoot you up a little. I was wondering when you would show up. Jesus is a good actor, isn't he?

"Come in and have some supper. I have some questions for you."

Chapter 5

Ransom

"THIS CAN BE EASY, OR this can be hard," Gonzales began. "I really don't care which. The result will be the same. You must be rich to have nothing better to do than to play around with your life like this. Someone will pay for you to live or you will die a slow and painful death.

"Bring the man in," he ordered.

"Senior Sanchez did not want to do the little things that we asked of him, and I want you to see the results." Gonzales continued. A limp figure with blackened eyes, bruises on his naked torso and a scalded back was brought in and thrown on the floor. Gonzales put his dirty boot on his back and ground his heel into the tortured flesh while Sanchez howled in pain. Then he kicked the nearly prostrate man in the testicles which elicited a heart-wrenching sobbing cry and caused him to ball up in a fetal position.

"Enough," Larry Jr. cried out. "You have made your point. I'll tell you who to contact."

"I see you are both wearing wedding rings," Gonzales observed. "How much do you think your wives would pay to get you back? Sometimes they don't, you know, and if so we'll do whatever we want with you."

"Both of us are newlyweds," Aron responded. "We are just starting out in life and don't have any real money."

"Somebody has some money somewhere or you would not be in a place like this," Gonzales insisted.

"Call my dad," Larry Jr., responded. "Maybe he can arrange something."

"Now we are getting somewhere," Gonzales responded. "Give me the number, and we will see what he has to say. I'll let him talk to you both. Take off your shoes and step outside."

Motioning to the figure rocking on the floor he told two men to take the weakly struggling man outside, and with guns drawn two others walked Larry Jr. and Aron across the porch and down to the ground surface.

"First a little demonstration, that I am serious," Gonzales said. He drew his pistol and shot Sanchez in the left leg, then the right leg and while he tried to drag away on his arms in the left arm and then the right arm as Larry Jr. and Aron looked on in horror.

Sanchez screamed and screamed. Gonzales reacted, "Quiet, can't you see I am making a telephone call," as we walked up and put a bullet in the struggling man's head.

"See, even I can show mercy," he quipped as he reloaded the Smith and Wesson revolver before replacing it in its holster.

Senora Sanchez ran out from the kitchen and threw her arms around her fallen husband.

Gonzales instructed four men to take the Sanchez family out into the desert in the pick-up and bury the body before bringing the others back, "Don't want him stinking up the place," he remarked.

"Call," Gonzales demanded. "Tell him nothing but that you and your compadre are all right. Then hand the phone to me."

"Larry Jr. took the flip phone and called a number. After six rings it answered.

"Austerhouse here," Larry Sr. answered.

"Dad this is Larry. Aron and I are all right, but there is someone here that wants to speak to you."

"Senior Austerhouse I am with your son and his compadre. After they saw a demonstration of the work I do they have agreed to my offer of a contract for $400,000 dollars for the job. You know how these banks in Mexico are. So that the work can be completed day after tomorrow, I need the cash dropped by sunrise Thursday or we will have to terminate the contract. Comprehende?"

"What the hell are you talking about," Larry Sr. demanded. "Put my son back on."

"I will be happy to let you hear from him," Gonzales said. "Un momento." Gonzales then put down the phone took a rifle from one of his men and with the butt slammed Larry Jr. across the kneecap.

Larry Jr. cried out and cursed, "You Bastard."

"I am sorry, your son seems to be temporally, how do you say, ur ... indisposed." Gonzales said in a flat tone. Perhaps you would like to hear from his friend? He will receive the same treatment."

"That is not necessary," Larry Sr. "I agree to your terms. It will take some time to gather the cash."

"You have until Thursday noon," Gonzales responded. "I'll call again Wednesday afternoon to confirm delivery. Things would go bad if news of our deal leaks out to anyone."

"I need to speak to Aron." Larry Sr. replied.

"Of course you may," Gonzales responded. "Here he is."

"Tell Sam that I love her and am all right," Aron blurted out into the out-stretched phone.

"Our business today is concluded," Gonzales said as he ended the call.

Samantha was in her office between meeting calls when her telephone lit up signaling an incoming call from an unfamiliar number. Picking up she was surprised to find that it was Larry Sr. on the line.

"Samantha, some difficulties have arisen concerning Aron's and Larry Jr.'s vacation plans. It seems that I am going to have to raise some cash to get them out of it by a Thursday deadline. I have talked to them both, and they are all right. You and the baby come down when you can. I will fill you in when you get here. We will manage this as a family affair."

Samantha declared a family crisis and postponed all of her meetings for a week and hurriedly got ready for her and Michael's trip to Texas. She and Aron had not managed to convert but a small portion of the gold coins to cash. Checking the balance in their reserve funds account, she found deposits totaling $25,000. Strongly suspecting that this was a kidnapping-ransom situation, she would take that much down with her just in case.

Albert found himself in another flying adventure with Larry Sr. rather earlier and longer than expected. The morning after Larry Sr. received the phone call they were on their way to Alpine. Small town banks don't keep large quantities of cash on hand, and a single withdrawal of $400,000 might deplete the money supply of every bank in town. He knew the bankers and the Sheriff from childhood, and both would need to be brought into the picture.

"Mr. Austerhouse, you sure you want to do this?" Albert questioned. "It's a long flight there and back."

"If it was your son that was at risk, wouldn't you?" Larry Sr. rebutted.

"Larry is my best friend, and I want him back too," Albert agreed. "You work the radio and just tell me what to do, and I will get us there until we are ready to land."

There was a tense meeting at the Alpine Savings and Loan bank between Austerhouse, Sheriff Randy Jones and Bank President Willard Foster.

"Willard, Randy, my son and son-in-law have been kidnapped by Mexican smugglers and are being held for a $400,000 ransom which is to be paid in cash by Thursday noon, "Larry Sr. began. "They want it in small used bills. They say, as usual, that they will kill them if I do not make the delivery on time or if I involve law enforcement."

"There is no time for the FBI or Texas Rangers anyway," Randy replied. "Whatever we do we will need to pull from local resources."

"Give me a couple of hours, and I will get the cash together," Willard added. "I don't know if it is my place to say this or not, but about half the time they kill the people anyway. Are you sure they are still alive?"

"I am to get another call Wednesday night, so at least they will keep them alive that long," Larry Sr. reported. "I also think I know about where they are. They are most likely at the old mine in the middle of The Solitario. We can't put anything out on the radio or telephone because they would likely intercept it or be informed. We'll mobilize out of my ranch at Study Butte."

Larry Jr. now limping badly on his right leg, was half drug into the old crusher building while Aron helped support his weak side. The machinery had been removed and the roof had collapsed decades ago. The debris had been cleared leaving a concrete floor. Wood had been piled in the middle of the room and the fire illuminated the interior.

"We are going to have a little entertainment tonight," Gonzales announced, taking a swig from a bottle of Mescal. "Senora Sanchez is going to be introduced to her new life as a prostitute, and you all are going to participate. Strip her and hold her down, he ordered. You, tall Gringo. You will be first."

Three men grabbed Aron, ripped off his Sarape and as he struggled duck-walked him to where he stood over the naked woman.

"I won't do this," he said.

"You know what happens to those that refuse my small request," Gonzales said as he grabbed a rifle and swinging it hard slammed it across his lower rib cage breaking ribs and causing Aron to fall to the floor.

"They are no fun," Gonzales commented. "Take them back and tie them up."

Grasping his sarape with one hand while holding his aching ribs with the other, Aron and Larry Jr. attempted to support each other as they hobbled away on bare feet to the sounds of screams and laughter emanating from the building.

Chapter 6

Rescue

At the Auserhause ranch Samantha started to list resources, just as she had for hundreds of meetings before. On a legal pad she drew three lines down from the top of the page: Manpower, Transportation and Equipment.

"Who do we have available to help that we can get here in 24 hours?" she questioned the group around the table. Gathered were Larry Sr. who was being wheeled around by Albert. Although the senior Auserhause had regained some of his former strength, his balance issues had been aggravated by the flight to Marfa.

Sheriff Randy Jones represented law enforcement. He was as lean and hard as a fence post and appeared as if he could drive nails with his fists. He was accompanied by Judy Alvarez, his communication specialist.

Joe Whitefeather, the Apache tribal Chief, was spokesman for indigenous interests. Dressed now in jeans and western shirt, his heritage was made manifest by his long black hair, black eyes and steady gaze. He was a somber figure at the poker table that his opponents found hard to read.

Although dressed for this occasion, The Rev. Billy Pederson, head of the Full Gospel Bible Camp and Nudist Colony came out of a since of responsibility since Aron and Larry Jr. had been participating in a camp-sponsored function.

The most flamboyantly dressed of the group was Pedro Avanza Salvaterra de la Granadelrio the descendent of the Spanish grantees of the property on which The Solitario was located. He was dressed in green trousers with silver button trim, a vest made of tooled leather with horn buttons worn over a shirt with voluminous sleeves and pointed patent-leather boots adorned with silver toe and heel caps. He looked like a character out of a "Three Musketeers" movie, but seriously real as a long thin dagger sticking out of his waistband attested.

As this was Larry Sr.'s house and he was one of those most directly impacted by the kidnapping, he spoke first. "The message that we want to send to this gang of thugs is that 'You don't mess with Texas,'" he began. "Whatever happens, I want to get the message across loud and clear. I have 10 hands who know this country and can ride and shoot."

On her tablet, Samantha listed 10 mounted ranch hands.

Sheriff Jones spoke next. "We have got to keep things legal. I will deputize everyone who participates, but there will be no killing unless the party is armed and resist arrest. I know that you might want to do more, but let's keep it legal. I do not want to call in the Rangers or Border Patrol until after we get your men back. They would just slow us down."

Samantha noted "one Sheriff" on her list and then motioned for Whitefeather to speak. "The tribes have a personal grudge against this gang. Twice they have kidnapped young girls, raped them and killed them or left them for dead. We have a group of young men, about 20, who are interested in learning the old ways and have been taught sign reading and tracking. This would be a real working experience for them, and we welcome the opportunity to put a stop to this gang."

"What shall I call them?" Samantha asked.

"You may call them braves," Whitefeather replied. "In this context 'braves' would be no insult."

Samantha noted 20 braves on her pad.

Reverent Billy spoke next. "I have the most professional group available. These are our security force which consist of eight men who served in the U.S. Special Forces, Navy Seals and Rangers. They came to us for solace and refuge. They have weapons, body armor and camo uniforms that match this terrane. They can move practically unseen, even in daylight. This is the group that can go into the camp, grab the two men and protect them until everyone else arrives."

"That's excellent," Samantha said as she noted eight-man extraction team.

"If I call now," Pedro Salvaterra began. "I can have 20 of the best horseman, ropers and whip users in the world come to help. They cannot bring guns with them, but if they are on a horse, no one is going to get away from them."

"Please call them," Samantha asked as she noted 20 horsemen on her pad.

"Do we know how many they are?" Samantha questioned.

Albert spoke up. "I wanted to learn something about the country before I came. A buddy of mine gave me passwords to access military grade satellite imagery. I looked when the satellite was over this morning. I counted 15 people around the old mine in the center of The Solitario. Two are maybe our guys, three were being drug out of a building, who may be a family being smuggled. It looks like there are about 10 gang members unless some were in the buildings all the time or on lookout. I tried infrared and came out with about the same number, but that was more difficult because animals give off heat too.

"We have them outnumbered," Samantha concluded.

"What do we have for transportation?"

"I have my Cessna," Larry Sr. said. "I have trucks and trailers to move my men and horses, but we have to do all of this at night, so they don't spot them or make it seem like a normal part of ranch operations – like rounding up cattle for branding. I can get my people there."

Reverent Billy said, "There are windmills out there that my men can go to and pretend to be servicing. The trucks will leave, but the men would stay and start to infiltrate into The Solitario from some unexpected approach – somewhere away from where the cattle are being worked.

"My caballeros, cowboys you call them, can help work the cattle and then go into an arroyo and rest their horses until it is time for them to ride in. Once they start it would be difficult to hide them.

"I can disperse the braves outside the rim to watch every goat trail that leads out of it," Whitefeather said. "Once the main part of the action is over, they could come in and find any bandits that might be attempting to hide."

"All that sounds good," Sheriff Jones said. "I will be where I can be most useful in coordinating things. In this business I think that things might go on that I do not want to see. Just remember, you do not kill unarmed individuals or those who give up – no matter how much you think they deserve it. You are sworn to uphold the laws of Texas."

"On the equipment side." Samantha continued. "Thanks to Albert we have two scans of their camp each day about two minutes long. Except to contact the kidnappers we want to stay off phones and radios. They are being constantly monitored by the Border Patrol, customs, Mexican Federales and who knows who else. When the insertion team has Aron and Larry, shoot a flare to signal that it is time for everyone to move in.

"The trick is to get most of them out of camp or at least distracted," Sheriff Jones interjected.

"We have just the thing to do that." Samantha said. "We are going to temp them out with the thing they want most - $400,000 in ransom money in small bills."

"How?" Larry Sr. asked and everyone listened with interest as Samantha outlined her plan. It was simple, effective, executable and comprehensive.

"Little lady, I'm impressed." Sherriff Jones commented. "If you weren't up there in Yankeeland, I'd hire you in a minute as Operations Officer.

"We'll have to move with whatever people we have when the time comes," Sheriff Jones concluded. "The more people we have the less likely the gang will offer serious resistance. All the riders need to be on their way tonight to possibly be ready to go by mid-morning day-after-tomorrow."

A cry from a crib interrupted the meeting. "I'm sorry gentlemen,

I have someone who needs a feed and wipe. Please get things going. I am going to busy for a while," Samantha said as she went to attend to her child.

An black rotary-dial telephone rang on the polished mesquite-topped table in the hall. The bone-white cholla-cactus legs trembled as Alfred rolled Larry Sr. to the phone.

"Austerhouse here," Larry Sr. answered.

"Have you received our present?" the voice asked.

"What present?" Austerhouse replied.

"Just a little something," the voice responded. "A small body part just to remind you that we are serious. Perhaps we may send another, larger part, tomorrow if we don't get our delivery on time."

Although Larry Sr.'s face was turning red with rage, he kept his phone voice down to a level that he would use if he were ordering horse feed. "Your delivery will be by air at 10:00 A.M. tomorrow morning. It is just as you asked, and I will drop it by myself."

"Excellent," the voice responded. "It will be expected, and if anything unexpected should occur you can expect a warm welcome."

"Lopez," Larry Sr. ordered. "Go to the post office and see if we have a package."

"Samantha," Larry Sr. called. "Those were the kidnappers. I don't want to worry you, but they have sent us something. 'A small body part' he said. I think that they may have cut one of Aron or Larry's fingers or toes. They did it even though they knew the money was coming. That is the kind of bastards they are. We have sworn not to kill them, but they may wish they were dead."

An hour later Lopez had returned with a small cardboard box and hat in hand brought it to his employer. "Senior Austerhouse, this was in your post office box, and it smells. I'm sorry."

"Everybody," Larry Sr. called. "It's here. Come and see."

Samantha came in along with Sherriff Jones and Albert.

"Samantha," Larry Sr. started with an unusual note of sympathy in his voice. "We need to know whose this is. We may have an injured man to get out. It could be Aron or Larry. I need you to identify it."

Retrieving a long-bladed ring-locked Spanish knife from his wheelchair side-pocket he cut the string on the package and unwrapped the cardboard box. Opening the box he found a knotted clear plastic bag which he held up. Inside was a blackened severed little toe.

"That's not Aron's" Samantha said. "It's too small."

"I do not know if this is Larry's or not," Larry Sr. stated.

"Let me look," Sherriff Jones asked. After he was handed the bag, he twisted the plastic tight around the toe and took it over to the light. Retrieving a 10-power magnifying glass from a vest pocket he looked at the toenail very closely. This toe belonged to a man – maybe one of those they were attempting to traffic into the U.S. They often kill the men and then traffic the women and children into the sex trade."

"That's horrible." Samantha replied. "While I am relieved that is not my husband's or Larry's we really need to put a stop to all of this. Is everything ready to go for tomorrow?"

"The roundup is going on, the pump repair crew has been sent, the caballeros are coming in and the braves will deploy as the sun sets." Sheriff Jones reported. "During the night they will move to their watch or assault positions."

"Mr. Austerhouse, is the Cessna ready?"

"Yes, it is all ready and the duffle bag that we are going to drop has been wired up to its parachute. That's made out of heavy canvas, grommeted all around and tied with nylon cord. We threw if off a windmill and it works. I put a bag of pesos in it so that it would fall right."

At 10:00 AM Aron and Larry heard the drone of an approaching aircraft. They had cloth sacks over their heads and could not see the plane. Sounding something like a large insect flying in a hatbox, the noise reverberated through the air.

Footfalls were heard as their captors left the shade of the mine buildings to watch their payday approach. The plane was at first obscured by the sun, but by shading their eyes they could see a prominent duffle bag tied to the struts of the aircraft on the co-pilot's side. A hand reached out the window, pulled a cord and the bag began to fall to be caught by a white parachute which opened. A cord attached to the zipper of the bag opened it and a flurry of money, like thousands of butterflies, started to drift towards the ground to fall a few hundred yards outside of camp.

Aron and Tom heard the shots of *"Dinero, dinero, alli, alli"* and sounds of running men as they ran towards their windfall profits. A door opened and slammed shut as their guard, not wanting to lose out on his share, hastened to join the others.

A few minutes later the door opened, and two others entered the room.

"It's all right. We are here to get you," Team Leader Neuville Blake said as he started to untie the pair. After his hands were free, Aron removed his hood. Even though he was inside a building, the light coming through the cracks in the walls and ceiling was such that he could hardly see.

"Your eyes will adjust slowly," Larry Jr. said. "When you step out into daylight you will not be able to see anything for a few minutes. We are going to have to let these guys lead us until we can see what's going on."

"Are you able to move," Neuville inquired?

"I am stiff and have broken ribs," Aron replied.

"You?" turning to Larry Jr.

"They hit me bad on one knee," Larry Jr. replied, "I can stumble around, but not at all well. I think the kneecap is busted."

"Where are the other captives?" Neuville asked. They are in the next building. There are three now. They killed the man days ago," Larry said. "They wanted to watch Aron take her too, but he refused, and they busted him with a gun butt and likely broke ribs. There is also a 10-year-old boy, and about a 4-year-old girl."

"Go," Neuville said to team member Pepe. "Check on them and keep them quiet as possible. If there is a guard, kill him. We need to get them all together so we can set up a perimeter."

Assisting Larry up he helped them out the door where the rest of the team stood at the ready. Seeing that the most substantial building was the stone crush house, he motioned the team to secure it while he gathered all of the captives. This was a position that he could defend.

"Pick up whatever weapons and ammo that you find and bring them with you," he ordered. Once inside the crush house, he stationed his men around the four walls of the building and fired a flair which rose into the sky and burst red.

Having either left or put down their gun in their rush to grab as many of the falling bills as possible, the men in the crater realized that something had gone very wrong as they saw the flair burst above them.

"Back to the mine," their leader Gonzales ordered. "They want your money. Get them."

While his men were returning to the mine, he found the duffel bag, cut the cords from it and with whatever money remained went in the opposite direction to where a helicopter might land near a feeding station. As he walked, he called to Mexico for a helicopter.

Riding over the rim along the road, the horsemen and the caballeros galloped towards the center of the caldera. Those rushing to assault the mine were stopped first by a spray of gunfire on the ground in front of them and later by the sound of approaching hooves which was at first a faint sound and then shook the desert air like thunder. Seeing the approaching riders moving to surround them, the gang members either threw up their hands in surrender or tried to escape with the money they had.

Moving figures in the brush were quickly spotted by the horseman and the *cabelleros* resorted to their lariats or whips to run down the individuals. The unlucky ones were lassoed and drug through the rocks and cactus until they stopped resisting.

Having returned to the ranch and refueled, Larry Sr. was once more prepared to go into the air with Albert by his side. Albert was armed with the AR-15 which now had an extended magazine.

"Albert," Larry Sr. said, "You are going to be my gunner. The only way that they could hope to escape is by flying in a helicopter from Mexico. We are going to hunt for it. I am going to fly by it close enough for you to shoot out its engine and bring it down. Son, are you really up to this?"

During the weeks of intimate contact with the old man, Albert, the orphan, felt that he had developed family bonds for the first time in his life. Initially these were having sex with Margaret and Elizabeth in their father's house and now by helping his best friend's father free his son from a deadly captor. He was ready to do whatever was asked of him to make sure that this gang leader was brought to justice.

"Yes sir," he said. "I'm ready." Albert replied, fully cognoscenti that for the first time he could remember someone had actually called him "Son."

The braves stationed along the rim around the caldera were wearing outfits that they used when they played extras in cowboy movies. Some real things that they carried were firearms dated from original trap-door Springfield rifles from the 1870s to 1873 Winchesters to bows and arrows. Some were replica guns, but the bows were newly made with arrow points fashioned from obsidian and salvaged steel. Those who captured or caused the surrender of one of the gang using the most primitive equipment received the greatest honor, and perhaps even an eagle feather.

Like hunters everywhere they awaited their prey. Some came. Frightened, sweating and exhausted by their rapid climb up the steep slopes, those gang members that made it that far were confronted by pointed rifles, bows and a "stop right there" order from one or more braves.

One attempted to draw a pistol, but was hit by an arrow in the shoulder. This shot was not fatal, but it was sufficient to cause him to drop his gun. The braves had asked Whitefeather if they could scalp their captives, since this was actually not killing them. But he had replied, "not even a little bit." In obedience to their orders, the captives were held down while bleach was applied to their scalps leaving a skunk-like stripe so their work would be instantly recognized.

Now over The Solitario, Larry Sr.'s suspicions were confirmed when a small Bell helicopter was seen taking off from near a stock tank.

"There he is," Albert replied excitedly. "Do I shoot from here?"

"I want to get you close enough to shoot out the engine. If they are killed when they crash, that's their bad luck. Once they cross the border we can't touch um."

The aerial battle was looking something like a dragon fly taking on a bumblebee. The Cessna, now 200 feet and behind the Bell helicopter begin its dive to approach from behind to pass on the pilot's side of the aircraft. With his hand on the Bell's control the pilot could not shoot back.

"Here goes," Larry Sr. said. "Aim from the tail to the engine and stop shooting when you reach the bubble. Put as many shots into it as you can. I am going to put you a little high so you might hit the rotors."

As fast as it took to describe it, the initial attack run was over. Albert fired as fast as he could shoot, but was unsure he hit anything except for hearing three pings of bullets against metal. He held on as Larry Sr. made a sharp bank to the left and swung around to reestablish his attack position.

Albert readjusted the rifle so that he could shoot more towards the front of the aircraft. In the meantime, the helicopter speeded up. Larry Sr. cut back on his throttle so that the speeds of the two aircraft were more nearly equal and Albert would be able to get more shots away.

As they approached for a second pass, the helicopter banked sharply to the right and shots were fired from the passenger's side of the helicopter. A ping was heard as a shot passed into the wing.

Albert attempted to adjust his shots, but got only two away with uncertain results as Larry Sr. turned the Cessna away from the helicopter.

"We will have to try for him at longer range," Larry Sr. observed. "Aim for the engine and shoot when you can. Did he hit us?"

"Yes, there is fuel coming from the right wing tank," Albert shouted.

"That's all right. I'll switch to the other tank, Larry Sr. replied as he manipulated the gauges.

"I think I got him that time," Albert reported after the fourth pass. There is smoke coming from the engine.

"Good. We should get him this time," Larry Sr. Said as he brought the aircraft closer.

Albert wiped his eyes and put his head down on the comb of the rifle. He lined up the iron sights and when they started to intersect the whirling circle of grey caused by the rotating tail rotor, he started pulling the trigger. One, two, three, four, five, six, seven, eight shots were fired before the aircraft separated.

As they were in a steep going-away climb, did not see the results of any bullet impacts. Then the helicopter started a spiral spin first up and then down as smoke billowed from beneath the main rotor.

"It's going down," Albert shouted. "We got it."

"Not quite yet," Larry Sr. replied. "We've got to get people to it somehow, and there is not much in the way of roads back here."

Flying over they saw the tail section was broken off and a column of black smoke was rising from the machine. There was one figures lying beside it.

Noting the location, Larry Sr. called Samantha, "They tried to escape in a helicopter, but it had an accident and crashed. It is about six miles northeast of the South Fork Ranch Road. There is one person on the ground. Maybe injured. I don't know. If Sherriff Jones is there tell him or call him. He can take care of it."

"Aron and Larry were injured," Samantha told the elder Austerhouse. "They are being taken to the hospital in Marfa. Me and the baby are leaving now. We got all the captives out. One of the gang was shot with an arrow, but will survive, and no one on either side was killed, so far as I know. Things are still happening, and the Border Patrol has been called. The Sheriff is dealing with them."

Chapter 7

Aftermath

Not even during the days that the mine and mill were in operation had so many people and vehicles descended on The Solitario. First to be loaded out were Pedro Avanza Salvaterra's caballeros who were started on their trip back to Mexico before the Border Patrol arrived.

Satisfied that no Texas laws had been violated by the posse that he needed to know about, Sherriff Jones wanted to send the braves and ranch hands home as soon as he had a sufficient number of deputies and patty wagons to handle the prisoners and had contact information on those willing to be involved. Most of his time was spent gathering preliminary statements about the details of the Sanchez's murder and the raping of his wife.

There were other bits of unfinished business that needed to be accomplished. The first was to disable Gonzales's cell phone so he could not call for more aid from Mexico. The next item was to locate the wrecked helicopter and capture Gonzales. Sheriff Jones dispatched six of Austerhouse'a ranch hands and two of the braves to look for them. They needed to leave immediately so they could spot any smoke that might still be emanating from the helicopter.

"This man is certainly armed and dangerous. You are to find, follow and harass him," Jones instructed. "If he holds-up somewhere, keep him there until we can get help to you. Water your horses here before you go. I don't think that there is much water out there. If he shoots at you, kill him."

There was also the matter of $400,000 dollars of U.S. currency that was blowing around the landscape or stuffed in the clothing of the smugglers. This task was assigned to the remaining braves who were promised a reward for what they found. Before any of the money could be taken away it had to be photographically documented to strengthen the case against the kidnappers.

Then there was also the problem of what to do with the peyote that the gang had gathered. These were the property of the landowner. Not only were they evidence, but they were used by several universities and pharmaceutical firms in investigations for end-of-life palliative care and treatment of mental disorders.

Because of the multiplicity of agencies that might be involved, care needed to be taken so the correct people were prosecuted for the right crimes, and that help and restitution were offered to the Sanchez family.

While their sarapes were useful for the purpose intended, there was also the need to find something for Larry and Aron to wear before being reintroduced into "polite society." Even after these needs were met, their clothes IDs, credit cards needed to be recovered from Reverent Pederson which would necessitate a stop by the Bible Camp in Terlingua.

Aron was not surprised that Samantha had largely coordinated the rescue efforts and was pleased beyond measure that she was at the Austerhouse ranch.

"Samantha, this is Aron," he blurted out over the Sheriff's phone. "Me and Larry Jr. are banged up with some broken bones, but are basically all right. They are going to take us to the hospital in Marfa. Everything went well, nobody was killed, but the lead smuggler is out in the desert somewhere. Have someone drive you to the hospital and meet us there."

"What happened?" Samantha asked excitedly.

"We'll have plenty of time to talk at the hospital," Aron responded. "I've got to give the phone to Larry Jr. so he can call Beth."

"Do that. Beth has been worried sick," Samantha informed her husband. "I've already told her that you both were safe, but she wants to hear from Larry Jr. She wants to know if she should come. Bye. Love you. Michael and I will see you soon."

"Beth, this is Larry," Larry Jr. began. "We're out. I'm beat up, but basically all right. Daddy's hands and friends came and got us, and most of the ransom money has been recovered."

"Well, I damn well hope that the days of you two playing Cowboys and Indians are over," she angrily replied. "That business like to have scared Sam and me to death. If I were the man and you were the woman, I'd keep you pregnant and tied to the kitchen stove for that kind of shit."

"I can help with that, you know," Larry Jr. responded.

"And what is this business about your daddy flying an air-combat mission against a drug lord?" she asked.

"He did? Did he?" Larry responded in amazement.

"Yes he did. He did the flying and Alfred was his gunner," Samantha affirmed. "They shot down that helicopter that was trying to escape to Mexico."

"That's my dad, and that's the family you married into," Larry Jr. proudly affirmed. "What can I say? Yes, we occasionally do that kind of stuff."

"Do you want me to come or not?" Beth demanded.

"We are going to be in the hospital in Marfa for a couple of days and then back to the ranch for maybe a week," Larry Jr. informed. "After we get out of the hospital would probably be the best time."

"I'll see when I can get there," Beth replied.

"They're loading us in the ambulance. I've got to go. Love you," Larry Jr. signed off.

"Love you too. You two damn well better take care of yourselves or I'll have you committed for that shit. There's no judge in North America who would contest the order," Beth concluded.

Slim Cooper, a veteran military policeman and among the younger of the Austerhouse ranch hands, was the first to spot the smoke from the downed helicopter. He dismounted, took his .308 Ruger Ranch Rifle out of its scabbard and climbed to a vantage point where he could see the helicopter with an apparently lifeless figure beside it.

He rejoined the group, appointed horse holders, and asked the others to cover him with their rifles as he and the two braves approached the site. He did not want the horses' and men's footprints obscuring any tracks that Gonzales might have made.

A Hispanic man, obviously the pilot, had been killed by a broken sliver of the aircraft's bubble being driven into his body. He had apparently been able to release himself from his seat belt and drag himself away from the aircraft before bleeding to death.

The fate of his passenger was more difficult to read. Gonzales had apparently avoided an immediately fatal injury, but a pool of blood on the ground on the other side of the aircraft indicated that he had been hurt. With a blood trail to follow, a dog could easily find him. Attempting to trail the armed fugitive through that rough country with its many potential ambush points seemed foolhardy.

Calling Sheriff Jones, Cooper gave his report. "We found the chopper. The pilot was killed during the crash, but Gonzales escaped. He was hurt enough to bleed, but I do not know how bad his injuries are. I don't want to risk anyone being killed going into the brush after him."

"Quite right," Sherriff Jones replied. "I'll send a team to photograph and extract the body. Stay there until the Rangers arrive, but keep people away from the helicopter. Guard your horses. He may try to steal one."

By the time the four Rangers arrived in two four-wheel drive vehicles it was already approaching dark. They removed the pilot's body and set up a camera overlooking the site. As darkness fell everyone left, leaving Gonzales to cope with the desert as best he could. Although they searched, they could not find the pilot's phone.

Before leaving the mine Sherriff Jones arranged that his canine search team be brought in to start as early as possible the next morning. "Tomorrow, we'll get him," he thought. He was counting on him being hurt, dehydrated and alone in a hostile environment. "But anything can happen out here. I had better not count my chickens too soon."

Samantha decided that she and Michael would be more comfortable at the ranch rather than waiting at the hospital. She delayed her departure for four hours to allow time for Aron and Larry Jr. to make the long trip out to Marfa. She was thankful that there were ambulances stationed at nearby fire houses and one did not have to come all the way from the hospital to get the injured men.

She was glad that Ricardo was driving, and the club cab Dodge had enough room for her and the baby. She was in the back seat with the baby seat and changing supplies while the cooler and baby formula had been placed in the front seat. She had a bottle warming between her legs, and soon it would be body temperature.

When they approached the cattle guard at the main road the driver stopped to talk to a young man who was carrying a gallon jug of water. At the end of their conversation he jumped into the back of the truck and threw in a battered military duffel bag.

"Do you know that person?" Samantha asked.

"No. He said that he would pay me 20 silver pesos to take him to the store where he could catch a bus," Ricardo replied.

When they arrived at the Study Butte Store, their passenger jumped off and waved as they departed. When they were out of sight Samantha retrieved her cell phone and called Sheriff Jones. "Sheriff Jones, Samantha here. We just dropped off a young man at the Study Butte store who was carrying a duffel bag who payed with the driver with silver pesos. I think that he may have been one of the bandits. He looked to be about 20, slim medium complexion, with dark hair. He was wearing jeans and a Doger's T-shirt."

"Thanks," Jones responded. "I will have someone pick him up. That is too much of a coincidence not to be somehow related to the kidnapping. Sam, never, ever, do that again. He would have been picked up by the border patrol, the bus or even one of my deputies. You never know who might be out there or what their intentions might be."

Going to the Emergency Room entrance to the Mercy Hospital in Marfa, Samantha wheeled the baby carriage in while Ricardo brought in the rest of Michael's supplies for what might be a long wait. It was already past dark, and she rang a buzzer to signal someone to come to the desk.

A nurse, dressed in green scrubs, walked in. "What can I do for you?" she asked.

"I am Samantha Goldfarb. My husband and brother-in-law were brought in earlier. They had been kidnapped."

"We are working on them now," the nurse replied. "They are being cleaned up, X-rayed and given IVs. I am sorry, but there is just not room in there for anyone but staff. They were given pain meds on their trip in and are groggy. Here are some forms for you to fill out. What you don't know leave blank. I'll bring you their billfolds and things. Most significantly, are they allergic to anything that you know of?"

"My husband, Aron, is not," Samantha affirmed. "I don't know about my brother-in-law. I'll call his wife and find out."

"I am going to move you into their room while you work on those forms," the nurse informed the increasingly flustered Samantha. "Someone from the floor will come for you as soon as they can. We still have COVID patents. Here are some masks for you to wear."

"What do you mean they won't let you see them? They are our husbands," Beth shouted over the phone.

"They said that as soon as they could move them out of the ER," Samantha informed her sister. "They have got to get them cleaned up first. I haven't seen the doctor yet."

What seemed like half-an-eternity later there was a rap on the door and two figures came in. Both were dressed in masks and scrubs. Blood stains could be seen on the arms of their scrubs and droplets of blood on their masks.

"Mrs. Goldfarb, I'm Doctor Breedlove. We have your husband and brother-in-law stabilized and their external wounds treated. The trip opened some of them. The X-rays show that your husband has two broken ribs which we are going to have to operate on. The other man's kneecap is broken, and we are going to have to replace it with an artificial cap. A specialist is going to come in tomorrow to do the surgery.

"They have so many smaller wounds and scrapes on their bodies that we have them soaking in a hydrogen peroxide solution in the hot tub in the rehab section of the hospital, and you and the baby will be able to go see them. They have been asking for you.

"After that we will bring them up and you can stay in the room with them."

"Are they going to be all right?" Samantha questioned.

"The real dangers are internal bleeding and secondary infections," Breedlove continued. "So far as we can tell there has been minimal bleeding. We'll take care of that when we operate in the morning.

"There is one other complication. They told us that they were forced to eat peyote. I don't know how this drug will cross-react with anything else including what he was given for pain on the trip here. I have people researching that now so we can select the safest anesthetic for their surgery tomorrow."

Samantha was taken to the rehab section of the hospital where a series of screens had been pulled around the hot tub. A nurse pulled a section away to allow Samantha and the baby carriage to enter. Aron and Larry Jr. were sitting about a foot apart on the shelf of the hot tub immersed up to their necks in a frothing boiling liquid which was already plastering a scum of dead skin cells and dirt on the sides of the tub.

Picking up the baby and walking up to the pair, Samantha commented to Michael, "This is your Dad and Uncle Larry, but they are a sorry looking bunch now."

"We feel like we have been put through a corn sheller and spit out for cobs," Larry Jr. said as he raised one arm to show the darkened, dirty and bruised skin.

"Look like it too," Samantha agreed.

"I feel like that too," Aron added. "I've got broken ribs and Larry has a busted kneecap. They tell us that it was you who came up with the plan to get us out of there."

"That was me," Samantha said. "If there is anything that I can do it is to organize things. I could not have done it alone, there were lots of other people involved – not all of which you know about. There were Sherriff Jones, Reverent Pederson's security team, hands from the ranch, caballeros from Mexico, Whitefeather's Apache braves, and later the Border Patrol, Rangers and Sherriff's deputies. The FBI and maybe the Food and Drug Administration may be involved too along with the Federal Aviation Agency. This is to say nothing of your friend Albert who flew your dad around and shot down the helicopter.

"They forced you to eat peyote? Are you all right from that?" Samantha questioned. "The doctor asked me about that, but I had never even heard of it."

"Yes, they did," Larry Jr. responded. "We were starving and thirsty. They fed it to us in a stew. We both tripped out on it. Mine was a pleasant, vacant —out-of-the-body type of thing. I really didn't care what happened to me. I can't say about Aron, I was so wrapped up in myself. I didn't really care about anybody or anything."

"Mine was something like that," Aron agreed. "My experiences were worse, something maybe related to the way my dad treated me when I was a kid. I felt worthless, but it did not seem to matter. I felt like there was something out there further that I could not quite reach, but was on the cusp of attaining. What that something was or how I could get there was not revealed. But I wanted to go there again. I can see how this stuff would be addictive.

"I think that I am over it now, except that maybe things don't bother me as much. Like maybe it is all small shit and not worth worrying about. I love you and Michael, and I really missed you. That has not changed."

The curtains were being pulled away and the nurses brought around some stainless steel devices with straps to hoist Aron and Larry out of the tub.

"I call the horse lifters," Larry Jr. informed. "They use them to help us get out of the tub."

"Mam, would you leave now?" the nurse asked. "Moving them is going to hurt, and there is nothing you can do to help."

Chapter 8

Recovery

"Where are we going to put all of these people?" Larry Sr. asked Helga. "We are going to have Larry Jr. and his wife; Aron, his wife and child; Samantha's dad; one of her half-sisters; a medical student from Detroit; and Aron's father and mother."

"I suppose we can separate them into family groups," Helga offered. "Larry Jr. and his wife, her father, and half-sister can stay at my parent's ranch and the others over here. The medical student can stay here to help look after Aron who had the more serious surgery and Albert can look after with Larry Jr. who wasn't as badly hurt."

"Wwe have gotten accustomed to Albert, and I had rather have him here," Larry Sr. interjected. "You never know when we might have to fly again."

Helga offered that "It would be good to set up a room for medical treatment and keep all of the supplies there. Both men are going to need their dressings changed and the baby looked after too. "We have a sink and commode in the old nursery and can rent a hospital bed to use when they change their dressings. Everyone is going to eat most of their meals here, so that can work."

"Larry Jr. along with Aron, Sam and their baby will be arriving tomorrow as will some of the others," Larry Sr. stated. "We'll plan on sleeping everyone here tonight. Sherriff Jones is going to drop by and give us an update. Then we will have our little fiesta and get everyone out of here. I'm glad to have them, but I will be glad to get them gone too."

"Don't you dare say such a thing," Helga admonished. "After what everyone has been through we need some time together."

It took two pick-ups to get Aron, Samantha, Michael and Larry Jr. back to the ranch. Both Aron and Larry were wheeled out to their rides and loaded into the vehicles. Larry with his cast was put in the passenger's front seat whereas Aron was put in the driver's side rear seat of the other vehicle and braced to sit sideways to cushion his ride as much as possible.

Although Samantha would have preferred to have ridden with her husband, there was more room in Larry Jr.'s truck for her and the baby's supplies. Albert drove Aron's vehicle and filled him in on his and Larry Sr.'s flying and shooting exploits while Samantha informed Larry Jr. about the details of their rescue.

This was the first time that Aron and Albert had the opportunity to become really acquainted although they knew many details of each other's lives.

"Albert, I know about your being an orphan, being Larry Jr.'s roommate in college, coming to Beth's and Larry Jr.'s wedding, likely giving COVID to his mother and dad and then coming down to help them recover. Then I understand you flew with Larry Sr. and shot down the helicopter that Gonzales was in.

"How are you fairing with all of this?" Aron asked.

"Aron, for the first time since my parents died, I feel like I am part of a family. I've got attached not only to Larry Jr.'s father and mother, but incidentally to you, Samantha and Michael. I know that we are not blood relatives, but I feel close to you as well. I don't really know how to explain it. I'm told that this is how men in a combat unit feel. How I don't know, but some kinds of bonds have developed between all of us."

"I know what you are talking about," Aron agreed. "The same kinds of things developed between us when we were trapped in our apartment in Detroit with the in-laws and movers. We became a sort of extended family.

"I need to warn you about my dad, David. He is super critical, tight with his money to a fault and much like Larry Sr. wants to have everything his way. Either these two are going to get along like long-lost friends or be at each-other's throats. I don't know which. Like with me, all of this is going to be radically different from anything he has ever experienced. I suspect that we are going to have to help my mother, Sarah, keep him calm and smooth things along."

"Thanks for the warning," Albert replied. "There are two houses. It looks like we need to put your mom and dad at his wife's place that is about ten miles from the ranch. Helga was thinking about something like that anyway, so things can work out. We need to give them some space and time to adjust to each other."

"Another person that will be coming is a Moslem Syrian refuge, Erderlan Pasha, who was about to do his hospital residency when the Civil War in Syria interrupted everything," Aron added. "He helped treat combat wounds before his family escaped to the U.S. So we will have you and him as medical professionals at the house."

"That will work," Albert agreed. "We've converted a nursery into a treatment room for you and Larry Jr. We can keep all the supplies and change your dressings and put in an exercise bike for you both. Although it's going to hurt at first, we've got to keep you all moving, including Larry Sr. Me and Mr. Pasha, can design treatment programs for everyone. That has been a real challenge with Larry Sr., but he is coming around."

"I bet," Aron agreed.

Once he got his leg adjusted in a reasonable position in the other truck, Larry Jr. had the opportunity to have a long talk about Beth with his wife's sister.

"Sam, I love Beth, I really do," Larry Jr. began. "There is a strain on our relationship with her all of a sudden blowing up over something or other. Has she always been like that?"

"For as long as I can remember," Samantha replied. "Even when we were children she had temper tantrums. She thought that maybe working with Special-Education kids would help her, but that seemed to only aggravate things. She has managed to control herself with the kids, but took her frustrations out on everyone else, which, I assume, now includes you."

"With my construction deadlines, COVID, workforce and supply problems, I've had my problems too," Larry Jr. related. "It's doubly bad when I come home and she takes off about something I had absolutely nothing to do with. I can't seem to do much about it. The sex-stuff helps, but that is a thin thread to hang a lifetime on when nothing else works."

"I sympathize with you Larry, but I don't know what to tell you," Samantha responded. "Even if she were to find other work, I see no reason that it would change these anger-management issues. The only thing I can suggest is therapy and maybe drug treatment, but she has always rejected those – she insists that there is nothing really wrong."

"Although it certainly wasn't planned for in a clinical situation," Larry Jr. revealed. "Aron and I both had recent experiences with peyote. Eating it seemed to mellow things out. I wonder if peyote might do Beth some good in helping her gain control of herself."

"There are a variety of research programs going on at the University Hospital, and Aron can maybe put you onto something about that," Samantha suggested. "My dad, Tom, did all sorts of drugs during the sixties, and I suppose that he had some experiences with peyote."

"I know peyote and where it grows," Larry Jr. informed. "But I don't want to take on something like that without clinical supervision. I don't know if Reverent Pederson and his group at the Bible Camp are using it or what the members of the Native American Church are doing, but I know the Indians have all-night ceremonies where they eat it. I can ask Albert to look into this and see what he can come up with."

After everyone arrived back at the ranch, a truck with a fresh driver did a turn-around to pick up the next arrivals at Alpine. They would not reach the ranch until after midnight. Beth, Tom and Margaret were scheduled to arrive from Atlanta; and Aron's parents and Erderlan's flight from Detroit would come tomorrow. Plans were made to lay out a 2:00 AM meal of cold cuts and salad for them.

Excited to be back at the ranch, but so stiff they could hardly move, Larry Jr. and Aron were half-carried into the ranch house and settled into chairs before supper. They declined drinks and settled for some pain pills. "I don't know how long we are going to last," Larry Jr. said. "Just something to eat, a bathroom and bed is about all I am good for now."

"Aron how do you feel?" Samantha asked.

"About the same," Aron replied. "I think that Larry Jr. and I are on the same schedule."

"Aron, let me take a quick look at your bandages," Albert requested. "I may need to change them. Larry, your cast should be all right. How did you do with crutches at the hospital?"

"Fine, I guess," Larry Jr. replied. "I made some trips up and down the hospital halls. It's getting up and down out of chairs and beds that gives me problems. It hurts, but I don't have any blood running down my leg. This cast is just damn clumsy, and I will be glad to get out of it."

"I've got the tools to cut you out of it, now that you are here," Albert informed. "When the stiches come out and your leg can get back to normal."

After supper, a bandage change and bed, Aron asked Samantha to put Michael in his arms. He could not lift him because of the pain in his side. He felt a since of home while he was holding his son. They weren't home, but they were together – something just a few days before he had reason to doubt would ever happen.

Larry Jr. decided to take a nap after supper and left instructions that he was to be woken up when his wife arrived. He was not the only one keeping the vigil. Albert was looking forward to seeing Margaret and perhaps renewing what he hoped might be a relationship. The fact that she had come hopefully indicated that she also wanted to see him. He did not know how things would develop, but at least he had a chance.

Although asleep in a chair in the living room, Aron woke to the sound of the door opening and Beth's voice calling, "Aron, where are you?"

Groggily responding, he replied, "Over here in the living room."

"I missed you so much," Beth replied as she approached, attempted to pull him to her and kissed him.

"Take it easy," Albert warned. "He hurts all over. You too can do stuff later, but right now we need to get everyone fed and to bed. I'll have some sandwiches, salads and fruit brought in for whoever wants them. I'll also fix anyone a nightcap who wants one."

"That is the best thing that I have heard all night," Tom replied. "I'll take a Bourbon and water, a ham sandwich with everything on it but hot peppers and a banana or something like that."

Albert picked up the pad by the phone and started taking orders. Margaret came to him, put a hand on his shoulder said, "I'll take ham too." and then whispered in his ear asked, "Are you also on the menu?"

Embarrassed at being approached so directly in front of everyone, Albert's face turned red, and he replied, "We'll work something out."

Tom walked over to his daughter and advised, "Let Albert do what he needs to do. I am sure that a place can be found for both of you somewhere."

The initial greetings and eatings taken care of, Albert was responsible for sorting out the night's sleeping arrangements. Larry Jr. and Beth were in one bedroom. Tom was put in another while he and Margaret would share his bedroom. It was 4:00 AM before he climbed into bed beside Margaret. They kissed, briefly embraced, made love and slept. Breakfast would be at 7:00 AM.

"I don't know which of you three boys look worse," Helga observed when she saw the three at the breakfast table. Although she had not known of Albert and Margaret's previous relationship, she quickly surmised what had occurred.

"Albert, you go back to bed. We'll manage without you until lunch. You will have a full day today once Aron's parents arrive."

Albert protested that he was all right, but Larry Sr. intervened with a strong, "Go back to bed. You are no good to us right now. We will need your best efforts over the next few days."

Margaret looked longingly at Albert as he left the table, but her father held up his hand indicating that she was not to follow. When she looked at him he mouthed "later," and she nodded her acceptance.

"You look like hell," was the greeting Aron got from his father when he, his mother and Erderlan arrived shortly before lunch.

"Actually, I am in better shape now than I was," Aron replied as he, on a pair of wooden crutches, negotiated a passenger through the screen door onto the porch.

Sarah ran to put her arms around her son, but Aron warned her off saying, "Sorry Mom, they have wired my ribs up, and it hurts to touch them."

"Let me help you get back inside," Erderlan said as he held the screen as Aron turned around and re-entered the living room.

A number of chairs and ottomans had been arranged so the group could sit in an irregular circle with the four convalescents arranged as comfortably as possible.

"Welcome to what has become the Austerhouse nursing home," Larry Sr. said. "We're glad to have our boys back, and happy to have you all here to help us celebrate the occasion. All of this is going to be fairly low key. Tomorrow night we'll have a band and invite everyone who participated in the rescue so you will have a chance to meet some of the finest people in Texas."

"I understand that you got most of the ransom money back?" David asked.

"Not exactly," Larry Sr. replied. "Most of it was picked up by the Apaches who helped capture the gang and some more from the gang members. The duffle bag was recovered and there was some money still in it. Sheriff Jones is going to bring it here tomorrow.

"Your son tells me that you are good with money. Tomorrow at our little fiesta the braves who gathered the money will be given a share of what they found. Twenty-five thousand will go back to Aron and Samantha to replace what they contributed to the ransom, and the remainder will go back to the bank.

"Will you be our paymaster and give it out?"

"I'll be happy to do that," David replied. "How much money are we talking about?"

"Something like $400,000," Larry Sr. replied.

"That is a lot of money to have lying around," David responded.

"That's true," David replied. "Men have been killed down here for far less. As soon as you count out what is owed, I'll make out a deposit slip and Sherriff Jones will take the rest back to the bank."

Samantha questioned, "Whatever happened to the gang leader, El Jefe Gonzales?"

"Sherriff Jones will fill us in when he comes," Larry Sr. responded.

"Let's get you settled in and we'll all meet back here again for supper. We're going to have some bar-b-que brisket, catfish and venison stew which will be among the things we will be cooking and serving tomorrow."

Erderlan stayed so that he could examine his patients while David and Sarah were being taken to the other ranch house.

"I am glad to meet you all," Erderlan began. "I have heard so much about you. I am sorry that everyone could not come. Keje and my boys, Bryan and Tom, would have loved it. Parts of Texas look like parts of Syria and Turkey where we have been. Please forgive me if I take a lot of pictures. They want to see and know about everything about their new country – particularly about real-life cowboys and Indians."

"When things settle down, you must bring them for a visit," Helga invited.

"Are you sure you want us to do this?" Erderlan questioned. "I have learned in America that invitations are often given, but no one really expects them to be accepted. I just got into med school, which is in part a repetition of what I have already done, but I will have time off during the summer. I would like to bring everyone down for a couple of weeks."

"I do miss having some kids around, It doesn't look like we are going to get any from Larry Jr. anytime soon, and yours are old enough to enjoy. Give us some dates and we will plan on your coming this summer," Helga affirmed.

"By that time I expect to be rid of this thing," Larry Sr. said pointing to his walker. "You and your family are welcome. One thing to warn you all about, turning so that he could look at Albert, Margaret and Tom, "This is a ranch and things breed around here – maybe something in the air or perhaps it's because of our bright stars. Whatever. Things happen. Unless you are really committed to each other, be careful. You have been warned."

"Thank you," Tom replied. "We'll just have to wait and see,"

This was not the first time that Albert had heard such frank sex talk, but repetition had not made him any more comfortable about it. He blushed, again, as Margaret led him away. The last remark that he heard was Larry Sr.'s statement, "The condoms are in the bedside table drawer," which was followed by a round of laughter.

Before supper the blast of a semi-horn was heard as a bob-tailed rig with Speedy Movers painted on the side appeared in the ranch yard. Inside was a burly white man dressed in coveralls, a younger black man with long dreadlocks dressed in a Shakespearian outfit and a mixed breed Lab-Shepard.

Hearing the commotion Alex looked out the window and saw Alex Polaskie, Henry Smith and Momma Dog dismounting from their vehicle.

"Erderlan, Alex and Henry are here. Help me out on the porch."

The two ranch dogs, Houston and Austin, regarded the newly arrived canine with suspicion, but after a round of nose and butt sniffing took on a more friendly stance and proceeded to show Momma Dog around the ranch.

"Momma Dog careful be," Henry warned. "For in strange surrounds you find yourself and animals see that never before you have smelt or seen that may do you kind or harm."

"Alex, Henry come up," Aron beckoned. "I can't get down the steps. I am so very glad that you could come."

"I am not going to miss a good party," Alex replied. I just happened to be nearby. I also brought you a delivery from Detroit."

This response elicited a laugh from Aron and Samantha who had joined Alex on the porch.

"And how is little Michael," Alex inquired.

"Big and growing like a tree," Samantha replied. "I see you brought Momma Dog too."

"Droopy Drawers insisted," Alex related. "She has played as big a part in this endeavor as anyone. The other dogs are showing her around the ranch. I can appreciate that because this is country that I have never seen either."

"Henry, I know what you are wearing, but why are you wearing it?" Aron asked.

"Sparse clothes do I have," Henry explained. "Tis better travel I as a gentleman from Verona than in my accustomed garb as 'Droopy Drawers.' A little velvet, a little lace and sometimes a rapier does this man dress complete."

"Do you know how to fight with sword?" Larry Sr. asked with a note of doubt in his voice.

"For stage fights we are taught," Henry explained. "I pray set up a candle on yon newel post, and I will demonstrate."

A candle was put in a single silver candle stick and placed on the newel post of the front steps of the ranch house. Henry took two steps back and turned away with his back facing the post. Pointing to Larry Jr. he said clap your hands when you are ready for me to start and keep clapping until I stop.

With the clap Henry spun on his feet while drawing his blade and lunged forward. With a right-to-left stroke he cut the top inch off the candle, before the second clap he had returned and made another cut with the back stroke. At the end of three claps the candle was in four pieces with only a quarter inch of stub remaining sticking above the silver candlestick. The movements of the blade were almost faster than the eye could follow.

"That was impressive," Larry Jr. observed as his father nodded in consent.

"When fight we must with swords and knives, we followers of the Bard's clever arts must our business know, else no one survives to do the next night's McBeth," Henry concluded.

Chapter 9

Fiesta

THE SUPPER GROUP HAD EXPANDED to include not only the Goldfarb and Austerhouse families but also Tom, Margaret, Albert, Erderlan, Alex and Henry. There were two empty chairs for Sherriff Jones and a deputy who were also expected. Shortly after everyone sat down the Sherriff arrived and he and Deputy Judy Alvarez hauled four copy paper boxes full of bills and coins into the house and placed them on the floor in the hall. After this task was accomplished the Sherriff and Deputy joined the diners.

"This is Sherriff Jones who can tell us what happened after Aron and Larry Jr. were taken to the hospital," Larry Sr. announced. "Please eat, and we will hear about that later."

After the meal was concluded and the dishes cleared away, Sherriff Jones began, "While the Apache braves and caballeros were rounding up the gang, I called in my deputies and paddy wagons to take the prisoners away. We accounted for everyone but 'El Jefe Gonzales,' as they called their boss. He had apparently escaped in a helicopter and several reported seeing it flying out.

"Somewhat later I got a call from Samantha saying that Larry Sr. had spotted a helicopter burning on the ground between The Solitario and the highway to the Big Bend Park headquarters in Presidio. I sent six ranch hands and two braves on horseback. They found the crashed chopper and the dead pilot beside it. They also found where Gonzales had pulled himself away.

"I ordered them to discontinue the search and bring the pilot's body back. Then I arranged for my tracking dog and team to go in the next morning.

"The next morning the tracking team found that Gonzales apparently broke a leg during the crash and was cut up. He apparently used the pilot's phone to call in another chopper and escaped. I have notified the Federales and state officials in Mexico, and they are looking for him. We want him on murder charges as well as kidnapping."

"Damn Bucky, I wish we could have pressed the issue that night and got him," Larry Sr. responded. "I don't like the idea that he is still on the loose. Hurt or not, he is still a very dangerous man."

"We did get his vehicles and his gang, and he is injured so that should cramp his style for a time," Sheriff Jones replied. "There is another matter. Those Federal boys, the FFA and the FBI, may force the issue if anything strange is found out about that helicopter crash. They are going to want to know why it was shot down and who did it.

"Larry Sr., Albert since you two found the helicopter you will be questioned. You can have a lawyer with you. I suggest that you get one. None of you should say anything about this event. This case may be heard by the FAA as well as state and federal grand juries. There is also the possibility that charges might be brought by Mexico since two of their nationals and one of their registered aircraft were involved.

"You can do whatever you want with the ransom money. We recovered all but about $1000. The missing money is out in the bushes somewhere."

The next day the mariachi band struck up their music at 4:00 PM to mark the official start of the fiesta. Components of the event had been arranged for their American guests including giving a delayed birthday party for Senora Sanchez's children Jesus and Maria.

Each group contributed something to the fiesta. Pedro Avanza Salvaterra de la Granadelrio and his caballeros were dressed in their holiday outfits and provided flashes of color as did the women. Salvaterra had brought a variety of Mexican beers and liquors with him. The Austerhouse ranch hands had been busily cooking beef, pork, goat and venison for the past two days to make taquitos, burritos and stews seasoned with a variety of ingredients. What was not consumed would be given away. The Apaches made corn fry bread seasoned with salt and native plants which they cooked hot and served on the spot. Although Alex and Henry had not planned on using the 10 Detroit deep dish pizzas they had brought down on dry ice these were put out as part of the fiesta.

Although not food, Reverent Pederson's contribution was to provide his armed eight-man team came prepared to provide security for the event. On their way in they concealed three security cameras at different points along the entrance road to the ranch. This was standard procedure when the colony held off-site events.

As the ranch was filling up with people, David who was counting the money in the office was feeling nervous. "This is too much money to have lying around with so many people coming and going that nobody knows," he told Larry Sr.

"That's a good point," Larry Sr. agreed. "Sheriff Jones, David has taken the money that he needs and made out the deposit slip. Please take the boxes and lock them in the trunk of your patrol car. I don't have room in my safe. I'll also have two of my men guard the door."

"Who's going to guard the guards?" David interjected. "I want to someone with me that I know."

"Who? Those two drivers? I can give them guns, but they might be more dangerous to themselves and everyone else than to any robbers," Larry Jr. suggested.

"As you saw, Henry knows now to use that sword," David recalled. "Have him sit with me while I count the money. He can disable someone with that sword before they could fire a shot. Please ask him to come."

"Good sir," Henry greeted. "Momma Dog and I will sit at thy right hand and will guard the proceeds as you direct with nose, ear and fang and blade. Nothing I warrant will unchallenged approach."

"Thank you," David replied with since of relief in his voice. "This amount of money would tempt almost anyone."

Although it was anticipated that the events might last quite late, the birthday part of the functions were held before much of the food was put out. Senora Sanchez and her children were being looked after at Pedro Salvaterra's hacienda. They would be returned to Mexico after the birthday party.

With appropriate music and song, the decorated *tres letches* cake was brought out where Jesus was expected to take the first bite out of it without using his hands. While he was bending over the cake his mother pushed his face into it while everyone else howled with laughter. He came up smiling and licking the sweet multicolored icing off his face. He was then ready to face his next challenge, breaking the *piñata* with a painted stick while blindfolded.

Jesus struck parts of the seven-pronged piñata but ultimately connected with the central part which caused it to break open and spill its load of candy and small toys on the ground which were gathered by Jesus, Maria and the ranch hands' children.

To his knowledge, David had never seen a Native American. Presented in a line before him were Apaches waiting to receive their portion of the ransom money they had recovered. David opened the envelopes, verified the count and had them sign a ledger that they had received the money. One by one they came to the door of the counting room. David sat behind a table at the door handing out envelopes for a payout of a bit over $40,000. When he gave out each envelope, he shook their hands and thanked them for the part they played in rescuing his son.

About half-way through the pay-out, Momma Dog rose to her feet, looked at a pair of French doors that opened onto the porch and growled. Alerted, Henry turned his attentions to the door and saw the handle move and then relax which was followed by footsteps running down the porch. Someone, perhaps with evil intent, had tried to come into the room.

The birthday part of the celebration concluded and the dancing had started while the remaining money was counted out to the ranch hands, caballeros and Mrs. Sanchez. "I feel better that my part of it is done," David remarked.

Although Henry had exchanged lines with many characters dressed in costume when on stage, he was taken aback when he saw Pedro Avanza Salvaterra de la Granadelrio striding towards him with the obvious intent of speaking to him. Managing a weak hello, he extended his hand.

Pedro Salvaterra stopped, looked him up and down and grasped Henry's wrist with his left hand took his right hand and shook it saying, "I have been told about you, and I appreciate a man who was has learned how to use a sword and knows something of the old ways of dress and honor."

"Pleased sir I am to your acquaintance make," Henry stammered. "How of service might I be?"

After a few seconds to make a mental translation and formulate his reply, Salvaterra responded, "You know of Zorro from the old TV stories. He was supposedly a Spanish Grandee who attempted to right some wrongs inflicted on the people by an oppressive colonial government. Period plays are presented about his supposed exploits, and writers, directors and instructors are needed to do them. Do you think that you might have a part in producing these?"

"Honored I am by your offer, but only a little, if any, part might I play," Henry explained. "The language of the period have I not. Sword fighting instructions must be rapid, precise and instantly understood. Perhaps, I could work with an actor who also speaks English for a few weeks, but that is the best I could manage."

"Jesus has been placed in my care," Salvaterra explained. "His schooling has been interrupted, and I was thinking that acting might be a useful skill for him to develop. Can you help?"

"Good sir," Henry responded. "Do you know what of you ask? This burden would be like a great stone. My own career would I be drowning so that he might swim with few chances than anyone would throw him, or me, a lifeline. Someday perhaps. Send him to school in

Mexico, and then aid might I be able to give, should he still have the desire. With regret, I could not at present, pursue this path."

A flashing red light on the dash of the Bible Camp's security van caught the attention of Team Leader Neuville. When he looked at the camera monitors, he quickly sought out Larry Sr. "Mr. Austerhouse, we may have a problem. Twenty men on motorcycles are coming in. Are you expecting anyone like that?"

Larry Sr. called Sheriff Jones over to see if he could identify them. "That looks like it might be *El Bastidos*, a motorcycle gang from Senora," Sheriff Jones speculated. "They are really bad actors. Close the gate and pull the vehicles around in a semi-circle behind the fence line. It's dark enough that we can blind them with our lights if they refuse to leave.

"Have your hands get their guns and take positions behind the cars. Give the caballeros and braves the ranch shotguns and whatever other guns they feel comfortable to shoot. Then send them to take positions along the road on the hillside on the other side of the fence. Bring everyone else into the house and put them down in the cellar. I declare everyone to be deputized."

Neuville who was watching the monitors, said "They are five miles out. Get everyone in position."

"What's happening?" Aron asked Samantha and she and Michael came into the dining room where he and Larry Jr. were sitting opposite each other.

"Sheriff Jones thinks that a Mexican motorcycle gang is coming to raid the ranch. They are going to confront them at the front gate," she replied as she helped support Michael to give him a bottle.

A large individual entered the room, stood opposite Larry Jr. and Aron and spoke, "Senior Austerhouse wants the Senora and Niño to go to the basement. The others are already there. You two may come whenever you can. They have set up barricades that you can shoot behind."

"I thought I knew everyone on the ranch. Who are you?" Larry Jr. questioned.

"I have been out in one of the line cabins," the man replied. "That is why you haven't seen me."

"Who's your boss?" Larry Sr. pressed.

Rather than reply, the man walked up to Aron and making a fist drove it into his damaged ribcage causing Aron to loudly gasp and then cry out in pain. He then took one of Aron crutches and banged Larry Jr. first across the head and then across his other knee.

Samantha screamed as the man pulled a knife and held it at Michael's throat.

"Be quiet, or he dies right here."

"We are going to walk out the front door together."

The intermitting flashing of lights and the rumble of the Harley engines were heard long before the first of bikes were seen. They were a combination of custom and factory bikes noted for their large size, aggressively cleated tires and heavy mud flaps. The gas tanks and fenders were painted with the green and yellow colors of the Mexican flag. Their helmets featured the eagle and snake engaged in deadly combat which foreshadowed their intent to any potential opponents. They slowed as they approached the fence and pulled up almost touching bike to bike.

Standing behind the open door of his old Ford Crown Victoria police car, Sheriff Jones ordered, "Stop right there. What brings you here?"

Allegro Enstranasos replied as he indicated a bandaged individual in the bike's sidecar, "We heard there was a fiesta, and we brought a business associate with some unfinished business to conduct to help you celebrate. It has been a long trip and we are thirsty."

"Bring up four cases of beer," Larry Sr. ordered. The riders shut off their engines, put their kick stands down and received the beer across the fence.

"Mr. Austerhouse it is good to meet you in person. You have something in the Sheriff's trunk that belongs to me," Gonzales stated with a coughing wheeze. "The bag is empty. Fill it please," he concluded as he handed a new duffle bag to Allegro to throw over the gate.

"The beer is good, but we have also incurred some expenses. A little remuneration and perhaps a parting gift would be appropriate – a little gasoline – a little money - a little food," Enstranasos suggested.

"How much?" Larry Sr. asked.

"Only a little," Enstranasos responded. "Say another $100,000 to provide security for your little party. I know you don't have the money here, so a few trucks, trailers and whatever we want to take from the house will do. You never know when isolated ranches like this might catch fire or who might show up these days. Perhaps you would like a little demonstration of what might happen?"

A scream came from behind the line of vehicles. A tough-looking man walked forward with Samantha in front of him. She was carrying a crying Michael and the man had a knife at her throat. Samantha put the bottle in the baby's mouth which quieted his protesting.

"Very well," Larry Sr. agreed. "We will pay you the money.

"Sheriff Jones give him the money."

Jones opened the trunk of his car, pulled out the three boxes and placed them on the ground behind the car.

"Open it," the knife-wielding man ordered.

As the kidnapper bent over to look inside the box, the knife moved a little away from Samantha's throat. Feeling his hand loosen its grip she broke away carrying Michael like a football.

Seeing an opening, Sheriff Jones stepped between them, grabbed the knife hand and pushed it back towards the kidnapper. As the pair struggled for the blade, Albert raised his rifle and put a bullet through the kidnapper's head dropping him instantly, but not before the Sheriff received a cut across the lower abdomen.

"Lights! Shoot!" the Sheriff shouted, and a general round of gunfire ensued as the motorcyclists, ranch hands and Apaches exchanged shots. With buckshot and bullets flying around them some of the gang members attempted to start their bikes and flee up the road.

Shots coming from the vicinity of the ranch house indicated that some of the gang members had infiltrated around the gate and were attempting to enter the house. Although hurting and stumbling from their injuries, Aron and Larry Jr. crawled through the hall into the living room. On their way Larry opened the drawers of the empty gun cabinet and found the only guns remaining were a cased pair of flintlock pistols which he loaded with .54-caliber patched round balls and 80 grains of FFFg black powder.

"These are the only guns that we have left. I'll load them up and prime them. To shoot them you pull the hammer back to full cock hold the gun on the target and pull the trigger," Larry Jr. instructed. "There will be a flash, and then the gun will fire. We'll set up in the kitchen."

As the two were making their way back to the kitchen a "woof" announced the presence of Henry and Momma Dog.

"Blessed to see you good sirs," Henry responded. "There are men in the dark outside and in maybe in the house. I don't know who is who. Alex and David with kitchen knives are guarding the basement door, and I with my blade and Momma Dog am defending the kitchen. Tom and Albert are at the gate."

"Block up the hall with chairs and that chopping block," Larry Jr. requested. Get Alex to help move that stove across the back door."

A spray of 9 mm. pistol rounds from a machine pistol peppered the outside of the building and broke a number of windowpanes. This was followed by footsteps on the porch.

A gun barrel poked through one of the panes. Larry Jr. lined up his pistol on an indistinct form outside and fired. The ancient flint from the White Hills of Dover that was napped by some nameless individual in the 1700s scraped against the steel, ignited the priming charge which flashed in the pan sending a jet of flame into the barrel and starting the heavy lead bullet on its way. Although choking smoke filled the room, the bullet was apparently effective as it was followed by the sounds of a body heavily falling on the porch and the pistol harmlessly discharging a round against the roof.

Larry Jr. rapidly reloaded by blowing down the barrel and then pouring an approximate charge down the bore which was followed by a ball and scrap of cloth. He then re-primed the gun.

The next assault was from the direction of the hall. A series of pistol bullets impacted against the chopping block as the sound was heard of chairs blocking the hall being thrown aside. Aron aligned his pistol so as to hit a person standing in the middle of the hall and fired his shot. A man cried out. Stumbling falling noises accompanied by the sounds of crashing china were heard as dishes were knocked from their shelves.

Henry saw a hand sticking through a window about to toss a grenade into the room, and he used his rapier to thrust at the figure behind the window. His blade found its target. The man dropped the grenade and Henry tossed it back out onto the porch. He heard the sound of men scrambling outside.

"Get down. Grenade!" he shouted.

The explosion blew out the two windows facing the porch and propelled shards of class into Aron and Larry Jr. None of these were debilitating wounds, but considerably added to their discomfort.

Henry jumped through the window, picked up a pistol and threw it inside. A man was rising to his feet and reaching for a pistol on the ground. Using one hand Henry vaulted over the porch railing and as his feet touched the ground drove his sword blade through the man's torso. Seeing no other targets in sight, he came back into the kitchen shouting, "It's me. Don't shoot."

As Henry came in he could see that the chopping block blocking the hall door had been pushed partly out of the way and a man brandishing a pistol was coming in.

Larry Jr., bracing his arm against the cast on his knee fired his hastily loaded shot and the man staggered back dropping his gun. Picking up the Glock 9mm., Aron fired four rapid shots down the dark hallway.

Truck lights started to illuminate the exterior of the ranch house as those who were defending the gate returned.

"You'll all right in there?" asked the welcome voice of Larry Sr.

"Yea, we are," Larry Jr. replied. "Be careful. Someone could still be in the house. Let the dogs in and search the house and outbuildings."

With Momma Dog on a leash, Henry and three armed ranch hands started their systematic search of the property after the bodies had been drug out into the yard. Houston picked up a trail of blood droplets which led to a tool shed. Austin followed with Momma Dog straining at her leash.

Seeing the animals approach a man made a run for it, but within a dozen steps surrounded by the three dogs. He reached to pull out a knife, but had his wrist grabbed by Momma Dog. As he tried to shake the dog off, Austin bit his legs and Houston grabbed a mouth full of jeans and butt and attempted to pull him to the ground.

"Drop the knife and be still," Henry ordered as he approached and brought the rapier to the man's throat. The man complied. With a whistle from one of the ranch hands Houston and Austin sat. Pointing toward the ranch house the man started walking with Houston and Austin on either side – just as if they were herding a stray.

"Anyone who is out there come to the ranch house and surrender," Sheriff Jones announced over the car's bullhorn. "We've got your bikes and there is nowhere to go. If you are wounded, cry out and we will get you."

"My security people can take care of things here for the next little while," Reverent Pederson stated. "You get that cut looked after. People have died from less than that."

"I've called in some help and ambulances to get this mess cleaned up," Sheriff Jones told the Reverent.

"Get Gonzales!" Larry Sr. shouted as he moved his walker closer to the gate. Under the lights he could see a figure apparently slumped over in the sidecar. "Put guns on him. He's playing opossum."

With car lights and flashlights illuminating the bike and sidecar, Enstranasos was seen slumped backwards and hanging to one side of the bike wide-eyed with a gaping open mouth. Gonzales was leaning forward over the front of the sidecar with his head down with both arms below the sidecar's rim. Now 15 feet from the side car, Larry Sr. drew a .44-40 Colt Peacemaker from a holster braced it on the frame of the walker, cocked it and pointed it towards Gonzales.

"Raise your hands where we can see them or I am going to end this right here," Larry Sr. demanded.

Gonzales did not move.

"Hold back. Let me do this," Slim Cooper suggested as he rode up on his horse. "I am going to put a rope on him." As Slim wound up to swing, Gonzales sat upright, and his hands came up with a grenade. Before Gonzales could throw the grenade, Slim's noose passed around his torso and when the horse pulled back the grenade fell to the bottom of the side cart. Gonzales had a look of terror on his face as he frantically reached for it, but a strong pull from the horse prevented it.

"Grenade. Everybody down," Slim shouted as he released the rope, turned the horse and spurred away.

Albert ran in front of Larry Sr. and bumped the walker. This caused the pistol to discharge its bullet through Albert's shirt nicking his side, striking the top of the side car and bouncing into Gonzales' throat. Albert fell on top of Larry Sr. and held him down when the grenade exploded, which sent fragments of the side cart and a red mist into the air. The motorbike burst into flames, and Gonzales waved in the air as he burned to death.

Tom and Lopez who had been standing by Samantha's side in back of the cars, rushed forward and grabbed the stunned Albert and Larry Sr. and took them back to the house.

Albert joined Sheriff Jones in the nursery where Erderlan was already working on the Sheriff.

Looking at the cut he evaluated the wound. "That cut penetrated the abdominal wall and has exposed some intestine, but fortunately did not penetrate it. I can clean it up, stitch you up and put a bandage on it or you can wait for an ambulance. "However, the ride could worsen the wound. I suggest that you let me do it and start you on some antibiotics. Once an infection in the gut gets started, it is very difficult to treat."

"Do what you need to do, Doc," Jones replied.

Albert, now somewhat recovered from the shock of the blast, raised his shirt and looked at the bullet hole. His skin showed a purple welt where the bullet had creased it, but no bleeding.

"Erderlan, I'm all right. I need to check on Larry Sr. and the others."

"You go ahead. There is not enough room in here. Why don't you see if anyone is still out on the road, then set up in the kitchen, and take care of the minor stuff there?"

"That sounds like a plan," Albert responded.

Albert found Larry Sr. in a recliner nursing a glass of George Dickel Rye. "Did I hurt you?" he asked.

"Better hurt that dead, which is what would have happened if you had not knocked you down. When the pistol went off, did I shoot you?"

"The bullet went through my shirt, but only touched me. I've got a bruise, but that's all," Albert responded.

"I guess we're even then about your puking on me," Larry Sr. ventured with a chuckle.

"Yea, I guess we are," Albert responded.

"You go see that young lady and show her that you are all right. She has been frantic about it. Do it now or she might never let you out of her sight again," Larry Jr. suggested.

"I will," Albert agreed.

The immediate crisis mostly passed, family groups were reuniting. David and Sarah emerged from the basement into the kitchen and saw the blasted-out windows, glass-strewn floor, and bullet holes in the walls which marked the desperate fight which had taken place a few minutes before.

"Aron, are you all right?" Sarah reflexively questioned.

"Yea Mom, I've got glass in my back and legs, a guy punched me in the ribs, my ears are ringing, I'm bloody all over and my hair is a mess," Aron enumerated. "Outside of that I'm just fine."

With that remark he tried to laugh, but it hurt too much which elicited laughter and relief from both parents.

"Aron, you did good," David responded with tears of happiness running down his face. "I'd hug you, but I know that would hurt too much."

"I'll take a rain check on that Dad," Aron responded with salty tears stinging the fresh scratches as they ran down his cheeks.

"Sam are you and the baby all right?" Aron asked as soon as his wife and son returned to the ranch.

"Just like another day at the office," Samantha replied, "A little hostility, a little fighting, a resolution of the situation and now the reports to do.

"Arron you look terrible. You need to get back to work. I don't know if you can stand anymore of this 'resting up.'"

After Albert with two ranch hands brought the last of the gang's wounded inside, he started organizing things in the kitchen. Margaret rushed up and put her arms around him, and he winced when she touched his bullet-burned side. "What's wrong? Are you hurt?" she asked.

"I've got a bullet burn on my side, but it's fine. I need to work on Aron and Larry Jr. in here. Can you help me clean this place up?" he entreated.

"Me and Larry Jr. are full of glass now. A grenade exploded outside and peppered us both. Fortunately, Henry and Momma Dog were behind the wall and didn't get hit. We just need to get this stuff out of us. We can't move without it cutting us," Aron interjected.

"I can't even sit down," Larry Jr. added.

Margaret, Sarah and Beth volunteered to do the glass picking while Tom, David, Henry and Alex worked with getting the bodies out of the house and continued to look for any stragglers that might be hiding in the brush or in the outbuildings. Instead of the scenic photos that Erderlan had planned, his camera was put to work documenting the scene as the ranch was slowly put back into operational condition.

Nearly eight hours later the gang members had been taken to jail, the injured were under guard in a wing at the hospital in Marfa, and the family members were once more gathered in the living room. After the night's events, they were so pumped full of adrenaline that they could not sleep.

"One thing that bothers me," Judy Alvarez stated, "is how did Gonzalez know that the money was in the Sheriff's car? Somebody had to tell him."

"Maybe it was that man that I fought with?" Sheriff Jones speculated.

"I don't think so," Larry Sr. replied. "No one here knew him. He came in with the gang."

"It had to be someone who was here as the fiesta started – someone here during the birthday party," Judy suggested. "Did anyone see anyone use the house telephone?"

"I did," Sarah replied. "It was the young man who had the birthday. He was talking very rapid Spanish to someone. I asked who he was calling, and he said that he was telling his aunt in Mexico about the party. I didn't think anything about it."

"That thug tortured, killed his father and mutilated his body. I saw it," Aron interjected. "Why would Jesus help Gonzales?"

"I'll tell you," a small voice erupted from a chair in the corner of the room where Jesus had pretended to be asleep. "I wondered how long it would take you to figure it out. I'll tell you everything. It is not to my advantage to attempt to hide things now."

"Jesus, you fed us; and then betrayed us after we tried to help you?" Larry Jr. questioned.

"Yes it was all an act," Jesus continued in a matter-of-fact way speaking in perfect English and in a manner that was beyond his years. "I learn quickly and easily. My father kept a small shop in Guatemala and wanted to get out of the country. He paid Gonzales to smuggle us to the U.S.

"Gonzales said that I reminded him of himself when he was a boy, but he was raised and abused in every way possible in a brothel. That was all the life he knew. He saw in me a way to get ahead by starting his own high-class brothel with young girls and boys and then blackmailing any important figures who came. He wanted me and my mother to help him run things, and my dad was just in the way, so he killed him.

"I had, or could learn, the computer skills to make this work; and the ransom money would be used to set everything up in one or more cities in Mexico.

"When we came to the mine he knew someone was there and sent me to find you and prod you into making a rescue attempt so you could be captured."

"But he beat you. I saw the marks," Aron said.

"Yes he did," Jesus replied. "He said only if I experienced pain could I really enjoy inflicting it on others. That was his life, and he was teaching me. He promised that we would have money, cars, women and anything else that we wanted. My mother and sister would be safe and made part of what he saw as his brothel empire.

"That is why I did what I did. That line of opportunities are gone now, and I am looking for new ways to provide for me, my mother and sister. Childhood has passed me by. I need education, and a legal way to make enough money to support us. Otherwise like Gonzales did, I will do what I must and likely meet a similar end. He was forced into that lifestyle. I don't want to be, but I will take it up if that is the only opportunity I have.

"Gentlemen and ladies this choice is yours, do you want me to be a businessman or a bandit? Going through the juvenile justice system is excellent training for a would-be gang leader. That is the default position. I had rather have my family looked after and be sent to a good school and college. You know my story. I am not going to run away. My life is literally in your hands. I am going to sleep now to await your decision."

Jesus walked out of the room and left those inside to debate his fate.

"That was quite a speech for a young man," Samantha remarked. "He is certainly intellectually gifted. He likely has some emotional, developmental and psychological problems; but he's smart."

"He's too smart," Sheriff Jones commented. "Anyone who is so callus as to have his father murdered and see this as just a practical business decision is too dangerous to put back in society until he gets straightened out. He is not an American national, and he is a juvenile. Outside of turning him over to immigration, there is nothing I can do with him. He may have been misguided, but what crime has he actually done outside of 'aiding and abiding in the commission of a felony.' He could put up a strong defense that Gonzales forced him to do what he did. I don't see that we could do anything with him here."

"I have taken him and his mother into my household," Pedro Avanza Salvaterra began. "Despite what he has done, I find that I like the young man. He may have been misguided, but I think that he is honest. He is certainly brilliant. I have the resources to guide him through what you would call high school, and by that time he can decide what he wants to master in college. With a mind like that it could be anything."

"In The Bard's tragedies heros like Othello are misled into doing a great wrong which they find impossible to once more set right," Henry commented. "Here the play of life offers the chance of not only recovery, but prosperity for all if Jesus is allowed to his full talents use. This boy, as a man, may the lives of millions boost to new heights. Even in a twisted way, Gonzales saw the brilliant star that he might become. Are we less than he? Can we not also see? I pray thee all. Let him have his chance and not into darkness cast."

"I can't take him back," Pedro Avanza Salvaterra said. "My *cabelleros* feel like he betrayed them. I think that they will kill him."

"We could enroll him in a military school and then he could possibly have an entryway into college. If people want to contribute to his upkeep, I'll handle the finances. He is a brilliant young man who could do much good in the world if given a chance," David suggested.

"If he can be put somewhere near us," Aron responded, "Sam and I could look after him. We've had good luck taking in strays.

THE *Goldfarb* CHRONICLES

Book 3

Brewster County Law

Chapter 1

Trial Day 1

LARRY SR. HAD A LONG discussion with his lawyer, son, Samantha and Albert as to whether he should plead guilty and stop the entire business of the three of them facing charges of premeditated murder. "After all, I'm kicking 80. If I go to prison for 10 years, I am not likely to make it out alive or in any condition to do anything more with my life. If the rest of you were not involved, I would just as soon that they hang me now. I would gladly swing on the gallows if they would let you two go."

"Nothing doing" Samantha responded. "We are in this together. The Prosecutor obviously considers your case as his best chance of winning a conviction. You are on home ground and are likely to have a sympathetic jury."

"I agree with Samantha.," Albert added. "I know you told me several times not to shoot to kill those people, and I did not. We wanted to stop them, not kill them. I think that the jury will understand that."

For several days the usually placid college town of Alpine, Texas, had been bustling with the arrival of reporters and TV vans representing not only the major U.S. networks but also those from Mexico and the U.K. The numbers of credentialed media had grown so large that they had to be mostly housed in one of the college auditoriums and the proceedings broadcast to a large screen. A lottery was to be held each evening to allow 10 members of the press to watch the proceedings in the courtroom.

"oyes, oyes, oyes," the Bailiff announced. "The Superior Court of Brewster County, Texas, is now in session. Judge Paul Prescott presiding. All rise."

Judge Prescott, a man of medium height with a striking head of white hair entered the chambers, took his seat, and gaveled the court to order. "You may be seated. I call The State of Texas vs. Larry Jason Austerhouse Sr. Is the prosecution ready to proceed?"

A young man in his late 20s, Murphy Brown, dressed in a fashionable business suit, rose from the prosecutor's table and replied, "We are Your Honor."

"Is the defense ready to proceed?"

"We are Your Honor," replied an older gentlemen in his 60s who was attired in a suite with a western cut with a pearl button shirt closed with a bolo tie. This was not the first time that Bucky Richards had appeared before Judge Prescott in the 40 years that he had practiced law in Brewster County. Because of the county's small population of less than 9,000 and huge geographic area, he had often had dealings with many of the inhabitants regarding land and water issues with occasional trials for more serious offenses. Although there were more than 20 possible charges that could have been brought against Austerhouse Sr. by both Federal and State agencies, he had managed to persuade all of the interests to drop the minor charges, such as giving flying lessons without a license, and violations of state park regulations in favor of one trial for Premeditated Murder to be held in Brewster County.

"I have some words for those visitors who are unfamiliar with this court," Judge Prescott began. "We are a sparsely populated, poor and huge county where it is a considerable inconvenience to empanel a jury. Out-of-town jury members will be housed in a Sull Ross college dormitory and all jurors will be sequestered there during deliberations.

"I expect the trial to be conducted speedily with the case going to the jury in four days. I understand that some salient facts have been stipulated by both parties to aid in this process. No photography will be allowed the court. Disable your cell phones now. Anyone whose cell phone rings will be immediately expelled. No demonstrations will be allowed the court.

"Breaks may be taken between testimonies when requested by the Jury, at mid-morning, for lunch and again at mid-afternoon."

"Will the attorneys approach the bench?"

"Mr. Brown, you were brought here from Dallas because our prosecutor had a conflict of interests over campaign contributions. Have you received adequate information from that office and the Sheriff's investigators to present your case?"

"I believe that I have, Your Honor," Brown affirmed.

"If you believe information is being willingly withheld, I will allow you some extra latitude to pursue it.

"Bucky is this understood?" Prescott concluded.

"That should be no problem. Nobody wants a mistrial," he affirmed.

After the lawyers had returned to their desks, Prescott announced, "The prosecutor may now present his opening statement."

"In a wanton and malicious disregard for the law, three individuals, Larry Austerhouse Sr., Albert Hopkins and Mrs. Samantha Goldfarb set into play a series of actions that resulted in the injury of one Mexican national Pepe Gonzales, known as El Jefe Gonzales, the murder of Bill Prospero and the destruction of a helicopter.

"We will show that Austerhouse with the aid of his son instructed Hopkins on how to use a powerful AR-15 rifle to shoot down the aircraft. We will further show that Austerhouse in violation of FFA regulations gave Hopkins flying lesions to enable them to commit the crime.

"We will also prove through a process of coercion and persuasion he caused Hopkins to kill three or more people a few days later to convert a one-time medical student pledged to save lives into a cold-blooded murderer eager to satisfy his blood lust.

"This will be the first of three trials – Austerhouse Sr. first, then Hopkins and finally that of Mrs. Samantha Goldfarb, the mastermind of the operation.

"The state will call witnesses to demonstrate motive, conspiracy, opportunity and execution."

"The defense may now present its opening statement," Prescott ordered.

"If it please the court," Richards began. "The defendant, Larry Austerhouse Sr., is being charged with an injury and murder resulting from his successful recovery of his son and son-in-law from a gang of Mexican kidnappers with the assistance of a Sheriff-led posse of his ranch hands, caballeros from Mexico, the guard detail from The Full Gospel Bible Camp and Nudist Colony of Terlingua and Apache Indians.

"This operation resulted in the successful liberation of his son and son-in-law, a Mexican woman and her two children whose husband had been murdered by El Jefe Gonzales and the capture of the entire gang. Only one gang member suffered an arrow wound and the only death on either side was the helicopter pilot. We will establish through evidence and testimony that these two deaths were an unintended accidental consequence of the operation.

"The rescue plan was conceived by Mrs. Samantha Goldfarb and others in the presence of Sheriff Jones and a deputy and agreed upon by all participants before the rescue was attempted.

"Seven days after the event when Larry Jr. and Aron Goldfarb were returned to the ranch to recover from injuries delivered on them by their kidnappers, the ranch was attacked by The *Bastidos* Motorcycle gang. This act was prompted by an injured Gonzales seeking revenge with the aid of an inside informant. This attack took place during a welcome-home fiesta given to thank those who had participated in the rescue. During this attack, Gonzales was killed after being shot at the ranch house gate by Larry Sr.

"Those attending the event, including Sheriff Jones, successfully repelled the attackers during an evening and night-time engagement which resulted in the killing, wounding and capturing of this notorious gang and the seizure of their motorbikes.

"Dependent on the approval of the Prosecuting Attorney, we are prepared to stipulate that these are the basic facts of the case."

"Both sides are cautioned that if you so stipulate that the events concerning the repulsion of the *Bastidos* which would not normally be allowed as testimony are now admissible which would include calling of witnesses and cross-examinations," Judge Prescott admonished.

"I do, Your Honor," Richards responded.

"Mr. Brown. Do you agree? I require an answer," Judge Prescott questioned.

"Sorry," Brown somewhat sheepishly answered, "I agree."

"Let the record show that both sides have agreed to the stipulations and that materials concerning the *Bastidos* raid may also be admitted," Judge Prescott intoned in a matter-of-fact voice.

"Mr. Brown. You may call your first witness."

"The prosecution calls Pedro Avanza Salvaterra de la Granadelrio," Brown announced.

The audience reacted with a gasp as the elaborately dressed Pedro Salvaterra walked down the aisle, was directed towards the witness chair and took his oath. He was wearing a silver fringed leather vest, a blue shirt, bloomy red trousers and patent leather riding boots.

"Thank you for coming," Brown began. "Do you remember a meeting at the Austerhouse Ranch on Tuesday before the attempt was made to rescue Larry Austerhouse Jr. and Aron Goldfarb from the kidnappers on Thursday?"

"I do," Salvaterra responded.

"Who was present at this meeting?" Brown asked.

"There was Samantha Goldfarb, Senior Austerhouse, Reverent Billy Pederson, the Apache leader Whitefeather, Sheriff Jones, a lady with him that I did not know and Austerhouse's foreman, Lopez, that I recall."

"Was Albert Hopkins also in the room?"

"Yes he was there too, but standing behind Senior Austerhouse and bringing things back and forth into the room like computer print-outs and images of The Solitario."

"Who was directing the meeting?"

"Mrs. Goldfarb was recording things on a pad, but everyone helped formulate the rescue plan.

"And what was the nature of this plan?"

"The plan was that the ransom money was to be put in a duffle bag and dropped just outside of the kidnapper's camp so that a rescue team could go in and free the captives while their guards were distracted."

"And what part was the defendant to play?"

"Senior Austerhouse was to fly his Cesena near the gang's camp and Albert was to throw the bag out of the aircraft allowing it to open on the way down and spill the ransom money out over the desert encouraging the men in the camp to go after it."

"Did Sheriff Jones approve of the plan?"

"He did. I remember his saying something to the effect that he would hire Mrs. Goldfarb as operations officer if she was available."

"Mr. Richards, you may now question the witness," Judge Prescott instructed."

"Senior Salvaterra, how would you characterize the tone of the meeting?"

"It was very businesslike."

"How in detail was the plan formulated?"

"Senora Goldfarb asked what assets each of the parties could bring, recorded those, discussed how they could be brought to the site, and everyone suggested how they might be used."

"Did anyone appear to be loud or argumentative during the process?"

"No. There was at times disagreement about details, and if such a complex plan could work. Everyone ultimately agreed. We all started pulling things together that night so everybody could be in place by Thursday morning."

"How many people were to be involved?"

"Approximately 60 I believe. There would be 30 riders on horseback between the ranch hands and my caballeros, 20 Apache braves and the six-man security team."

"Who would be armed?"

"The security team had ARs and modern weapons, the ranch hands their saddle rifles and pistols, the braves had historic weapons from western movies and my caballeros had their lariats and whips."

"Thanks you. That is all the questions that I have."

"Mr. Prosecutor, do you have any questions on redirect?" Judge Prescott asked.

"I do Your Honor. During what you described as a business-like discussion, did Larry Austerhouse Sr. make any statement?"

"I remember him saying something like 'You don't mess with Texas' or something to that effect."

"Did he appear to be angry when he said this?"

"Not as I recall. This was delivered in a very matter-of-fact voice."

"I have no further questions of this witness."

"Mr. Richards do you have any follow-up questions?" Judge Prescott asked.

"None, Your Honor."

"The witness may step down. Does the Prosecutor or Defense see any need for this witness to stay for the remainder of the trial?"

Both sides agreed that no further testimony would be required, and that Salvaterra could be dismissed.

"Senior Salvaterra, thank you for your testimony. You are dismissed from these proceedings. We will break for lunch and resume here at 1:30."

At a private dining room at the Holland Hotel the defense team gathered and discussed the trial.

"How do you think things are going?" Larry Sr. asked Bucky Richards.

"It is too early to say much of anything. I think that we have a good jury panel. We have some teachers, merchants, retirees and a professor from Sul Ross and as many as I could select from other parts of the county that might be more sympathetic to your case."

"The prosecutor is going to harp on that 'You don't mess with Texas,' statement to indicate that you had the intention of killing Gonzales all along. Otherwise, his testimony was completely neutral."

"Who will he call next?" Larry Jr. inquired.

"I have a list of possible witnesses, but they are not necessarily in the order that they will be called," Richards responded pensively. "Ultimately he will call Sheriff Jones, but he will likely be among the last witnesses."

At the resumption of the trial Brown called Chief Whitefeather to the stand.

"Chief Whitefeather were you present at a meeting held at the Austerhouse ranch the night before Aron Goldfarb and Larry Austerhouse were rescued?"

"I was."

"If it please the court, to expedite the proceedings I will ask the clerk to read the previous testimony to the witness and ask if he agrees as to the people present, the general tone of the meeting and the statement made by the defendant, and ask follow-up questions about that statement?" Brown requested.

"Counsels will approach the bench," Judge Prescott ordered. "Does the defense agree?"

"Yes, with the provision that my rights to cross-examine any aspect of both testimonies is not restricted," Richard assented.

"Very well Mr. Brown, you may continue with your witness" Judge Peterson ordered.

"Mr. Whitefeather, did you or your tribe have any reasons to be involved in this rescue attempt?" Brown began.

"The tribe had previous run-ins with the El Jefe Gonzales gang including intimidation, rape and possible murders. I was eager to cooperate in any effort to bring him to justice."

"Did you have any tribal members who were prepared to offer useful services to the effort?"

"We had some 20 young men who were being trained in the old ways of how to track and survive in the desert. They often acted as extras in Western movies and wanted to participate."

"How were they to be used?"

"They were to be deployed on the paths leading out of The Solitario to intercept any gang members who might attempt to slip away and track any who fled."

"How were they armed?"

"They were armed with single-shot and lever-action rifles from the 1800s, bows, arrows and lances."

"How were they to use these weapons?"

"They were instructed not to shoot unless someone shot at them or attempted to pull a gun."

"Were these instructions followed?"

"They were. One gang member attempted to draw a pistol and was shot in the shoulder by an arrow. He, I understand, was the only one wounded in the operation."

"At the meeting at the Austerhouse ranch did you hear the defendant make the statement to the effect of 'not to mess with Texas.'"

"Yes, I heard something to that effect. I think it was, 'I want to show them not to mess with Texas.'"

"Your witness, Mr. Richards."

"Mr. Whitefeather, when the 'not to mess with Texas,' statement was made, how was that delivered?"

"I'm sorry. I don't know what you mean?"

"Was Mr. Austerhouse shouting, or enraged or otherwise emotional when he made it?"

"No. No he was not. He spoke in a normal speaking voice. The same that he used throughout the meeting."

"Thank you, Whitefeather. I have no further questions for this witness."

Judge Prescott looked at his watch and announced, "One more witness, Mr. Brown."

"The prosecution calls The Reverent Billy Pederson."

"Reverent Peterson. You were also at this meeting at the Austerhouse ranch were you not?"

"I was."

"Why did you attend?"

"I am the head of the Full Gospel Bible Camp and Nudist Colony and sponsored the week-long survival experience at The Solitario that Larry Austerhouse Jr. and Aron Goldfarb were participating in. I felt a since of responsibility to get them back unharmed."

"Others who testified had people and assets available to participate in the rescue. Did the Bible Camp have any such assets?"

"Yes, we did. We have a six-man volunteer security team which consist of former Seals and Rangers that we use to safeguard our activities, since as nudists we would not carry arms and would be an easy target for anyone who might wish to do us harm."

"How was this team to participate in the rescue?"

"They were to sneak in close to the camp and while the gang was trying to gather the ransom money, they were to free the men and the captive family and keep them safe until the horsemen could arrive."

"How were these men armed?"

"They were carrying AR-15 style rifles, Ruger Ranch semi-auto rifles, various handguns and knives. Team Leader Neuville can give you more information about the guns. We needed to be at least as well armed as any of the drug gangs in the border area."

"Did you hear Mr. Austerhouse make any statements at the meeting."

"He described how he and Albert could use his Cessna and drop the money near the camp."

"Do you recall anything else he might have said?"

"Not outside of the general nature of conversation when he welcomed people and so on."

"Did he say that he was going to do anything in regards to the gang members."

"Not that I remember. I was more impressed that at 80, recovering from COVID-19 and using a walker he and Albert were flying that Cessna around. I had serious reservations about whether he could really do it until he told me that he and Albert had already taken longer flights, including to Alpine to get the ransom money."

"Your witness, Mr. Richards," Brown concluded.

"No questions, Your Honor."

"Very well gentlemen. This session is concluded. The jury is admonished that they are not to look at any TV coverage of the trial, discuss the case among themselves, read any newspapers, check their phones, or e-mails, or talk about the case with family members. In case of family emergencies information may be relayed through the Clerk of the Court.

"This court will reconvene at 9:00 AM. Remain seated while the jury is escorted from the courthouse."

With a slam of the gavel, the first day of the trial was concluded.

"Can he call us to testify?" was the first question that Samantha asked Bucky Richards while they were having supper in the private dining room at the hotel."

"Yes, he can. Even though you and Albert are also facing charges for the same incident, you may be called to testify at this trial. You may legitimately take the 5,th if you feel that the reply will be harmful to your own defense. You and Albert are both on his witness list. I don't know if it will be tomorrow, but it might be."

Chapter 2

Trial Day 2

MILD TEMPERATURES WELCOMED THE ATTENDEES as they walked into the courthouse to resume the trial. Although most had seen the two-story red-brick courthouse most of their lives, the low angle sun illuminated the red brick and the white trim around the windows. The long windows had been opened to allow as much breeze as possible to circulate through the crowded courtroom.

Following the usual call to order by the Judge and seating of the jury, Prosecutor Brown called Neuville Blake to the stand and resumed his questioning.

"Neuville Blake you are responsible for the security arrangements for the Full Gospel Bible Camp and Nudist colony of Terlingua, are you not?"

"I have held that position for five years," Blake affirmed.

"Previously you were a Navy Seal. Is that correct?"

"I was an active team leader, armorer and ultimately retired with the rank of Captain."

"In these positions, you acquired training and became knowledgeable of modern military firearms."

"That is correct. I have examined, shot and used most types of domestic and foreign military weapons."

"Due to your training and wide experience, may we consider you an expert on the use and capabilities of AR-15 type rifles."

"I don't call myself that, but yes I have worked with many of the AR-type rifles."

"Your Honor, I would like to submit that Neuville Blake is an expert in AR-type rifles and request that his testimony as to the use and capability of these guns be allowed in this court."

"Does the defense accept the witness as an expert in this area?" Judge Prescott asked.

"We do, Your Honor so long as the questions are confined to the mechanics and capability of the firearm," Richards responded.

"On the evidence table in front of you is a rifle. Would you examine it and describe it to the members of the Jury."

Neuville went over to the table, picked up the rifle, worked the action and examined the markings on the gun. "This is a Daniel Defense DDSVI semi-automatic rifle in .308 Winchester caliber. This is a .30-caliber version of the AR15 which was first made in the 5.56 NATO cartridge which is a .22 caliber."

"Could you explain the difference between semi-automatic and fully automatic firearms?"

"Fully automatic guns, fire continuously as long as the trigger is held down. The semi-autos may be fired only one shot at the time."

"Which is the more powerful round - the .308 or the .22 caliber?"

"The .308 has a heavier bullet and offers greater range and penetration on hard targets than the .22."

"What is the .308 used for?"

"It was developed as the 7.62 NATO cartridge for use in the M-14 military rifle and M-60 machine gun and became a very popular deer hunting caliber in a variety of guns ranging from single-shots, bolt-actions, pumps to semiautomatic rifles from most of the world's gun makers."

"Why did the U.S. Military go to a smaller, less powerful cartridge, when it adopted the AR-15 as its battle rifle?"

"The smaller-caliber ARs were much easier to control when shot fully automatic than the M-14s, and in the close-range battle conditions of the Vietnam War it provided the likelihood of more hits on targets per round fired. In addition, the cartridge was lighter in weight and more rounds could be carried into combat."

"At one time these rifles and large capacity magazines were banned because they were claimed to have no legitimate sporting use and were the preferred tools of mass murderers. Do you recall seeing such discussions in the media?"

"Objection, Your Honor. Council is introducing prejudicial information in order to inflame the passions of the jury."

"Sustained," Judge Prescott responded.

"Have these and similar rifles with large-capacity magazines been used in mass murders across the country?"

"Objection, Your Honor to this entire line of questioning."

"Sustained. Either move on to another topic or dismiss the witness," Judge Prescott ordered.

"Is the powerful .30-caliber round fired from this rifle capable of disabling a helicopter?"

"It depends on where the round hits. It will penetrate the thin skin of an aircraft and can damage control wires, engines and internal parts. There is a lot of empty space in a helicopter, and it is possible for a round to pass between the rotors of the aircraft without hitting a vital part."

"That is correct," Blake confirmed.

"Your witness, Mr. Richards."

"Mr. Blake," Richards began, "is it true that you have AR-15 style rifles in your security force?"

"Yes it is."

"Why did you select such guns?"

"My men were already well trained with these guns, they were readily available on the commercial gun market and they provided a firepower advantage if we were to be engaged in a serious firefight as we were during the raid on the ranch."

"Do these guns have legitimate sporting uses?"

"Objection Your Honor. I do not see the significance of this line of questions." Brown interjected.

"Sustained." Judge Prescott responded.

"Mr. Neuville, are you implying that because of the relatively large target area presented by the pilot of a helicopter the surest way to bring down a helicopter is to kill the pilot."

"Objection your honor, Council is leading the witness," Brown interjected.

"Sustained. Bucky you know better than this. I will have no more such interjections," Judge Prescott warned.

"No further questions," Richards concluded.

"Your Honor my next witness is Larry Austerhouse Junior who because of recent surgery cannot sit in the witness chair. May it please the court that he be allowed to testify from a wheelchair?"

Judge Prescott agreed and Larry Jr. wheeled himself in and positioned himself facing the jury in front of the witness chair. After he was sworn in, Brown began. "How did you know Albert Hopkins?"

"We were roommates at Tulane for four years. I was working on my E.E. degree and he was a pre-med student."

"You became close friends?"

"We did."

"How did he come to live at the Austerhous ranch?"

"My parents contracted COVID-19 at my wedding. Because medical school was closed, he agreed to come to the ranch and give them physical therapy while they recovered."

"While you were at Tulane, did you notice that he was unusually interested in guns or violent movies?"

"No. Quite to the contrary. He worked part time in the ER rooms at hospitals in New Orleans and commonly saw gunshot victims of all ages brought in. He knew I hunted and grew up with guns, but that was not his experience."

"When he came to the ranch did your father encourage him to learn how to kill people?"

"Objection Your Honor. Council is attempting to prejudice the jury."

"Sustained," Judge Prescott ordered.

"Did your father encourage you to teach him how to shoot?"

"Not particularly, I don't think. Dad wanted me to show him about life in West Texas, and that included shooting among many other things."

"You did take him shooting?"

"I took him and my brother-in-law Aron out to the range, where we shot a variety of guns, including that rifle. I wanted him to become comfortable with them so we could go hunting while he was here as well as to know how to defend himself if it became necessary."

"How did he do?"

"For a first-time shooter he did reasonably well. I wanted him to practice more so I could be confident that he could take deer at 100-yards with the rifle. I had planned on his doing more shooting before deer season."

"Was this when he first expressed a desire to kill people?"

"No. Not then. Not ever."

"You had trained him to use a rifle specifically made to kill people and he expressed no desire to see how it worked for its designed purpose?"

"Absolutely not! He wanted to hunt with me, and expressed interested in the rifle from that point of view and how small a group he could shoot, but for no other reasons."

"Did you practice rapid fire with the gun?"

"I did show him how to use a sling to steady the gun and we shot a couple of rapid-fire strings."

"Approximately how tight a group did he shoot rapid fire at 50 yards?"

"Ten shots in about a 6-inch group from a rest, as I recall."

"When did you hear about Hopkins and your father flying together?"

"When I arrived at the ranch, he said that Dad had persuaded him to help him get into the Cessna and after he took off showed him how to keep the aircraft on a heading. He did this as part of Dad's physical therapy along with horseback riding and other exercises."

"Were there later flying events?"

"There were. Once to Marfa to pick up the ransom money, one to drop the money at the kidnapper's camp and I later learned of a third flight where he spotted the helicopter."

"Did you witness the preparations for that flight?"

"No. I did not. I was held captive at The Solitario along with Aron."

"During the later raid on the ranch by the *Bastidos* did you witness Hopkins doing any shooting?"

"No. I was in the house with Aron where we helped fight off an assault. After the event Albert told me he killed one of the gang who had gotten into the house and came out with a knife at Samantha's throat. Samantha said that she managed to duck down, and he put a bullet into the man's head. He said that he also shot into the group of men and motorcycles in front of the gate, but he did not know if he hit anyone or not in the confusion of the firefight."

"Did he gloat over the men that he shot?"

"No. He said that he along with Erderlan Pasha gave first aid to those who were wounded. Between the two of them they likely saved the lives of several of the criminals."

"He told you this, but you did not personally witness these events, is that correct?"

"That is correct."

"He bragged about this later, no doubt?"

"Not that I ever heard about. He was remorseful about having killed anyone. He did so to save the lives of those he knew, including a mother with an infant."

"Is it possible you think that he did this to embellish his own image and impress others, possibly his girlfriend Margaret Williams who was also in the house at the time?"

"Objection," Richards said as he rose to his feet. "The witness cannot possibly know the motivations of another person."

"No further questions at this time. Your witness Counselor."

"Mr. Austerhouse, do you remember how or why that particular Daniel Defense AR came to be among the guns at the ranch?" Richards inquired.

"That was a Christmas gift from my Dad. He bought two of them - one for himself and one for me."

"Did he tell you why he purchased them?"

"He wanted to have them around in case the ranch was ever attacked, as it was, and also to use for hunting coyotes, wild hogs and deer. These were high-precision firearms and he enjoyed shooting them."

"Did anything about his buying these guns strike you as being unusual or unexpected?"

"No. This was just buying another gun. This was just something that he did from time to time."

"How many guns had your father accumulated?"

"I don't really know, perhaps 50, maybe more. Some were antiques, others were modern."

"No further questions."

"Mr. Austerhouse, was Hopkins paid or in any way compensated for his services? Richards asked.

"No he was not. I paid for his tickets to Alpine, but except for his room and board he received no compensation."

"Are we, and this court, to believe that he did this out of the goodness of his own heart, rather than getting anything out of it?"

"Albert felt responsible for giving my Dad and Mother COVID, and I think he came partly out of an obligation he felt to help them recover. Another reason that because the college was closed, he had nowhere else to stay until his medical school reopened. It was a mutually beneficial arrangement."

"No further questions."

The prosecution calls Blair Adams.

"Mr. Adams, you are the ranch mechanic who is responsible for keeping all of the machinery around the ranch working?"

"I am."

"How long have you held that position?"

"About 10 years, I suppose."

"Among your duties, were you trained to keep the Cessna serviced and operational?"

"Yes, that was among my duties. I had factory training on that aircraft."

"On the day that the rescue operation took place did you supervise the fueling and take off of the aircraft?"

"I did."

"Could you describe the aircraft being loaded for the first flight?"

"It was somewhat early, about 8:00 AM. There was Mr. Austerhouse and Mr. Hopkins. Hopkins and I helped Mr. Austerhouse get into the pilot's seat. Then we attached the duffle bag to the wing struts and tied it so that it would be released by pulling on a rope. We had a problem with water in the carburetor. I cleared that in a few minutes. The engine was run up, the chocks removed and the aircraft took off normally into about a six-mile-an-hour wind."

"On this first flight did you see any firearms loaded into the aircraft?"

"No. I did not. Only the passengers and the duffle bag.

"What happened when they returned?"

"Some 20-minutes later they landed back on the strip. They shut down, and I topped off the fuel, which did not take much. Mr. Austerhouse remained in the seat while Hopkins went to the car and came back with an AR-type rifle which looked like the one on the table. He climbed back into the Cessna with the rifle, they started up and flew off again."

"When did they return?"

"This was a somewhat longer trip – perhaps about 30 minutes this time."

"Did they tell you what they had done?"

"Mr. Austerhouse told me that they had seen a downed burning helicopter and radioed the information to Samantha to relay to the Sheriff. They asked me to use the land-line and notify the airport tower in Alpine that a crashed helicopter had been spotted, and a man was lying beside it."

"What happened after that?"

"They left, and I checked out the Cessna and put it back into the hanger."

"During your post-flight inspection did you see any bullet holes in the aircraft?"

"Yes there was one in the wing that penetrated into the gas tank. I had to purge the tank, and repair it the next day. There were lead fragments in the bottom of the tank from a bullet."

"No more questions. Your witness."

"Was anything said to you about why this second flight took place or what they intended to do?" Richards asked.

"When they got back from their second trip, they told me that from what they could see from the air the kidnappers had been rounded up, and that both Aron and Larry Jr. waved at them. Mr. Austerhouse and Albert were happy about that. I suppose that they wanted to see what was going on."

"Could such bullets have disabled the aircraft or set it afire."

"The Cessna is not built to repel gunfire. They could if they happened to hit in the right places or hit the pilot."

"No further questions."

"Mr. Brown I see you have three more witnesses. You may call your next witness now and the other two after lunch," Judge Prescott directed.

The Prosecution calls Ansel Slipp.

Slipp approached to be sworn in. He was in his 50s dressed in a somewhat worn grey suit with a brightly colored paisley-patterned tie.

"Mr. Slipp, Brown began. "You are an accident investigator for the Federal Aviation Agency are you not?"

"I am."

"Did you conduct the investigation of the recent helicopter crash in Brewster County near a topographic feature called The Solitario?"

"That is correct."

"Would you describe the general condition of the wreck?"

"The aircraft was a small two-man Bell helicopter which was mostly intact. The tail section was broken, and the prominent clear plastic bubble was detached and resting away from the aircraft. There was a comparatively small amount of fire damage at the top of the engine, but the gas tank did not ignite."

"Were you able to determine the cause of the crash?"

"There were a number of bullet holes in the tail of the aircraft and at least one hit on the engine and perhaps others on the rotors of the aircraft. The aircraft was downed by gunfire."

"Were any bullets recovered that were sufficiently intact to determine the gun that they were fired from."

"Unfortunately, they were not. One bullet was recovered from the engine. It was a 30-caliber full metal patched bullet and had a weight of about 130 grains."

"Was this bullet weight and style consistent with bullets that might be fired from a 7.62 AR-15 rifle like that one on the witness table?"

"Objection, Your Honor. "While we accept this witness as an expert on matters of aviation, we do not accept his qualifications in matters of bullet identification and ballistics."

"Sustained," Judge Prescott ruled.

"What was the diameter of the holes found in the aircraft's skin?"

"These were approximately .30-caliber."

"Is this the same diameter as bullets that might be fired from rifles using the .308 Winchester and 7.62 NATO cartridges used in the rifle on the evidence table."

"Objection," Richards interjected.

"Sustained," Judge Prescott repeated. "Let's move on."

"Was this aircraft shot down by gunfire which caused catastrophic failure of its mechanical components and the crash?"

"Yes. The helicopter was definitely shot down."

"Your witness."

"Mr. Slipp concerning the distribution of bullet hits on the aircraft, were there any hits on the bubble area of the helicopter?" Richards began.

"None that I could find."

"Were there any bullet hits on the helicopter's seats?"

"None that I could find."

"If someone were going to make a concerted effort to kill the pilot or passenger by shooting down the helicopter from another aircraft won't the men in the exposed bubble offer a larger and easier target?"

"Objection," Brown stated.

"Overruled," Judge Prescott responded.

"They would offer a larger target," Slipp agreed.

"No more questions."

"We will reconvene at 1:30 PM," Judge Prescott responded. "Please notify Mrs. Goldfarb and Sherriff Jones that they will testify this afternoon. Mr. Brown you will be expected to rest your case to allow equal time for the Defense. Please remain seated while the jury is escorted to lunch."

"I'm worried," Sam told the assembled family at the dinner table. "What if Michael raises a fuss while I am on the stand?"

"Don't worry, I will take him out if he needs a change or anything," Helga replied.

"Having him on the stand with you will provoke a sympathetic response from the jury," Richards responded. "Even if the judge orders Michael out of the courtroom, that will help. Just tell the truth as well as you can remember it. If something unexpected comes up, I can call you back as a defense witness. If we tell it exactly as it was, I think we can swing the jury."

"Don't worry about me," Larry Sr. added. "However things turn out I am going to be all right. It's you young folks with the rest of your lives ahead of you that I am concerned about."

"The Prosecution calls Mrs. Samantha Goldfarb."

Samantha with the child in her arms walked up to the witness chair, and took her seat in the witness chair while Michael, squirming to find a comfortable position, looked at his new surroundings.

"Madam will that child be quiet during these proceedings?" Judge Prescott asked.

"I do not know Your Honor," Sam replied.

"Can someone take him away if he raises a fuss?"

"Yes, Your Honor. The Defense is prepared for that contingency," Richards replied.

"Mr. Brown, you may proceed."

"Mrs. Goldfarb previous testimony from several witnesses relates that you were in attendance during a Tuesday meeting at the Austerhouse ranch before the rescue of your husband and Larry Austerhouse Jr. Is that correct?"

"Yes. It is."

"What was your roll at this meeting?"

"I acted as coordinator and administrator."

"What exactly did you do?"

"I listed the assets, recorded the contributions that each of the participants could make, discussed the transportation to the site, made up a schedule of events, and helped formulate the over-all plan for the rescue."

"Is it true that you were largely responsible for the formulation of the rescue plan?"

"Largely responsible might be too strong a word, but I did help put it together in a form that everyone could agree to."

"Was Sheriff Jones a participant in this meeting?"

"Yes. He was."

"Did he specifically approve of the final plan?"

"Yes. He did."

"What was to be his actions the day of the rescue."

"He was going to remain at the ranch to coordinate efforts while the rescue was taking place and then go there for on-the-ground supervision for the mop-up and recovery operations."

"Do you remember him specifically approving of Larry Austerhouse's and Albert Hopkins' flight to drop the money near the kidnapper's camp?"

"Yes. I do."

"Did he know of and approve of the second flight Austerhouse was to make that day?"

"That was not discussed during that meeting."

"To the best of your knowledge, did he know of and approve of a second flight the next day while the rescue was taking place?"

"Not that I witnessed. After the drop was made and the rescue started, he called his deputies and told them to drive to The Solitario. Who he might have contacted on the road, I do not know."

"Did you receive a radio message from Mr. Austerhouse after Sheriff Jones left?"

"Yes. Mr. Austerhouse radioed to Blair that he had spotted a smoking crashed helicopter with a man lying beside it and for me to contact the Sheriff and inform him. He also said that he had flown over the area and that the kidnappers were being rounded up and that my husband and Larry Jr. had apparently been safely rescued."

"When you contacted the Sheriff did he confirm this information?"

"He did."

"I take it from your testimony that the aircraft's radio messages could be received at the ranch, but not directly by the Sheriff."

"That is correct. Blair was monitoring the aircraft's radio frequency and telling me what was going on."

"Could this relay system have worked in reverse? Could a response from the Sheriff been relayed to Austerhouse Senior in the aircraft by your radio set at the ranch?"

"Not directly. Blair and I could listen through speakers in the house, but could not send a message. The transmitter was at the hanger."

"Would it have been possible for Austerhouse Senior to have called the ranch to ask you to relay a message to Sheriff Jones to ask for permission to shoot down the helicopter and receive a response back before the helicopter crossed the Mexican border?"

"Objection Your Honor," Richards interjected. "There are too many possible variables that the witness could not possibly know. The position and speed of the helicopter would be one factor. Did the Sheriff have his radio on? Was radio reception good that particular time, and so on?"

"Objection sustained," Judge Prescott announced.

"At any time in your presence or hearing did Sheriff Jones authorize the shooting down of the escaping helicopter?"

"Not to my knowledge. No."

"No further questions."

"Mrs. Goldfarb during that meeting at the ranch before the rescue did Sheriff Jones say anything about how the posse of some 60-odd people was to be sworn in?" Brown questioned.

"He said that he would make them deputies and that they were sworn to uphold the laws of Texas."

"How in particular was this done?"

"We, all of us in the room, including Albert Hopkins, took that oath."

"How was the oath to be transmitted to the others who might only appear at the capture site the next day?"

"The oath was printed on paper, Xerox copies were made, and these were cut into cards, dated for the following day and signed by the Sheriff with a space for the signature of the participants."

"Did Larry Austerhouse Sr. and Albert Hopkins receive such cards?"

"They did. I saw both Mr. Austerhouse and Albert sign them and put them into their billfolds. The others were to be distributed to the security squad, ranch hands, caballeros and Apaches the next morning."

"Did you conspire with Austerhouse Sr., Albert Hopkins or anyone else the night before the shooting or the day after to have any aircraft flying out of the Solitario that might be carrying the kidnappers shot down?"

"I did not."

"Did you hear of or know of any such plan from any source?"

"I did not."

"Could such a plan been formulated and executed by Larry Austerhouse Sr. and Albert Hopkins without your knowledge?"

"Yes. It could."

"Did in fact, Larry Austerhouse Sr. and Albert Hopkins in violation of their sworn oaths execute a plan to find and shoot down any helicopter attempting to escape from The Solitario and murder its occupants?"

"Objection, Your Honor. Counsel is asking the witness to offer a conclusion on the guilt of the accused."

"Sustained. Counsel Brown you are to confine your questions to matters of fact, and not ask witnesses questions based on suppositions or conjectural evidence. I have warned you once about this. If you persist

I will have you thrown out of the court and declare a mistrial. Is that clear?" Judge Prescott demanded.

"Yes. Your Honor," Brown replied.

"No further questions. Your witness, Mr. Roberts."

"Mrs. Goldfarb," Richards began. "At no time did you discuss or help plan with Larry Austerhous Senior or with anyone else that the Cesena make a second flight to The Solitario during that Tuesday meeting or the day of the rescue."

"No I did not."

"Did you at any time express the desire that Gonzales or anyone else be killed?"

"No. I did not."

"Did you encourage Albert Hopkins to shoot anyone?"

"No. I did not."

"Did you thank him for potentially saving your life and that of your son Michael."

"Yes. I did."

"How."

"I hugged and kissed him."

"How did he react?"

"He appeared surprised, turned red and tried to pull away."

"No further questions, Your Honor."

"Mr. Brown, do you have any questions on redirect."

"Mrs. Goldfarb. You did not want the man who kidnapped and brutalized your husband and brother-in-law dead?" Brown queried.

"No. I just wanted Aron and Larry Jr. back safe. Whatever happened to El Jefe Gonzales, I was perfectly happy to leave in the hands of the courts."

"Did you ever hear Larry Austerhouse Sr. express any opinion on what should happen to Gonzales?"

"Only something about teaching him 'not to mess with Texas,' but this was just a passing comment."

"No further questions at this time, Brown concluded."

"Mr. Brown please call your last witness." Judge Prescott ordered.

"Sheriff Beauregard Jones to the stand please."

Rather than his usual working outfit, Sheriff Jones was in fancy western dress with snakeskin boots, pinstriped brown trousers, a silver concha pistol belt, vest, shirt with pearl buttons and a bolo tie. Those visiting journalists wanted a show, and he was prepared to give them one.

"Sheriff Jones, how long have you been the Sheriff of Brewster County?"

"Sixteen years."

"During that time have you ever had contact with the defendant?"

"We grew up together and have known each other since our Grammar School days."

"Would it be a fair assumption to make that you are friends?"

"Yes. It would."

"As you are the Sheriff and he is one of the landowners in the county, have you and he ever had professional relations?"

"Over the years there have been cases of trespass, cattle rustling and theft of fuel that I can remember. He has also been a county commissioner and been called for jury duty at least once that I recall."

"Has he ever been, to your knowledge, been previously charged with a crime either in Brewster County or elsewhere?"

"In preparation for this case I did an FBI case scan on Mr. Austerhouse which covers all 50 states, and could find no indications of previous criminal behavior."

"One must run for Sheriff in Texas counties. Did Austerhouse Sr. ever give you campaign contributions?"

"Several times."

"What was the amount of these contributions?"

"As I recall usually from $500 to $1000."

"How did this compare with your usual contribution?"

"They were larger. Most of the money I received was $10 or $20 donations."

"For these larger contributions did Austerhouse Sr. ever ask you for anything?"

"I did receive occasional invitations to social gatherings at the house or elsewhere, but these were because I was a friend, not because he wanted a favor."

"Not even for a traffic ticket?"

"No. Not even for that."

"Turning to the meeting at the Austerhouse ranch the night before the rescue attempt was made. Were you at this meeting?"

"I was."

"Who would you say was running the meeting?"

"I can't say that any one person was, but Mrs. Goldfarb was collecting information and more or less organizing things."

"Did she suggest the plan that was ultimately used? That is the plan to drop the money in the desert and while the kidnappers were looking for it sent the Security Team in to rescue and protect them until the horsemen could arrive?"

"That is correct."

"Except for you and your deputy, this plan involved no law-enforcement officers such as the FBI, Rangers or even your county Sheriff's Department members. Why not?"

"The kidnappers may have been tipped off about our rescue attempt if these other agencies or even my own people were involved. I knew that I had at least one informant in my organization and the drug and smuggler gangs have others. It is difficult to keep anything secret for very long. This is the reason that I used local resources and got them on site as soon as possible. This close to the border almost all radio and telephone communications are monitored by someone, either legally or illegally. If the gang found out we were coming, I have no doubt that they would have killed the men immediately."

"Once you arrived, would you say that the operation was a success?"

"Mostly it was. Aron Goldfarb and Larry Austerhouse Jr. were rescued, but both had injuries. The gang members had surrendered and most of the ransom money had been recovered. The leader of the gang El Jefe Gonzales, although injured, called in another chopper and got away."

"Were there any aircraft reported or seen by you in the vicinity?"

"Men told me that they had heard and seen a small Bell helicopter fly into The Solitario and set down briefly and fly out again. Austerhous' Cessna returned and circled the mine area. Aron Goldfarb and Larry Jr. waved at them and he wagged his wings in response. He then flew away to the south."

"That would be towards the Mexican border?"

"Yes, it would."

"That was the same direction that the helicopter took when it escaped?"

"I was told that it had flown off in that direction, but I did not see the event."

"You could not directly contact the aircraft?"

"No. My radios were on different frequencies, and we could not communicate directly."

"Did you make any attempt to tell him to follow and shoot down the aircraft?"

"I did not."

"Did you at the meeting or at any other time encourage or order Larry Austerhouse Senior to pursue and shoot down that helicopter."

"No. I did not."

"Was the possibility that El Jefe might attempt to escape by helicopter mentioned at that meeting or later?"

"No. It was not."

"Were Larry Austerhouse Senior and Albert Hopkins acting without orders from you and on their own initiative pursuing that helicopter?"

"They were."

"Did you at any time give them orders to shoot that aircraft down if they had the opportunity?"

"I did not."

"Did you give your posse of which Austerhouse and Hopkins were sworn members instructions on the use of deadly force?"

"I made it clear that these arrests were to be done under the laws of Texas and that deadly force could only be employed to save their lives or the lives of others."

"If they had shot down the helicopter and subsequently killed those inside it, would this be in violation of your expressed orders?"

"Objection Your Honor," Richards said as he rose from the table. This is in fact the subject matter of this trial. The Prosecution is wanting the Sheriff to act as both judge and jury."

"Sustained," Judge Prescott ordered. "Brown you are on very shaky ground here. Reformulate your question."

"Very well Your Honor. Sheriff Jones did you at any time order Larry Austerhouse Senior and Albert Hopkins to pursue and shoot down that helicopter?"

"Objection, Your Honor. The witness has already answered this question."

"Sustained, Judge Presscott rulled with a degree of firmness in his voice.

"No more questions of this witness, Your Honor."

"Your Honor. Sheriff Jones is an important witness for the Defense. I prefer to defer my examination of this witness until tomorrow morning when the witness and the jury are rested."

"Is the Prosecution prepared to rest its case at this time?"

"We are Your Honor. The Prosecution rest."

"We will resume at 9:00 AM. This court is adjourned. Please remain seated while the jury is escorted out of the building."

Chapter 3

Trial Day 3

"**S**HERIFF JONES, YOU ARE REMINDED that you are still under oath,"

"Yes, Your Honor," Sheriff Jones replied.

"Sheriff Jones," Richards began, "in your posse you had some 65 people at several different locations. There were the ranch hands and caballeros staging at the western entrance to The Solitario. Some six miles from them the security team from the Bible Camp was sneaking in from a windmill site. Scattered around the rim Apache Braves were deployed on exit trails, and Austerhouse Senior was flying in with his Cessna. Could you directly supervise all aspects of this operation?"

"No. I could not be in all of those places at once."

"Were you in radio or other contact with these groups?"

"No. I was not."

"If you could not be there, how did they know what to do?"

"There was a previously agreed upon plan. That after the kidnappers left the camp to gather the money the security team would secure the hostages and shoot a flare when they had them in a secure place until those riding in from outside the rim could arrive."

"Everyone did not have radios or phones so they could be in constant communication with each other during the operation?"

"Many had phones, but because of the danger of the kidnappers intercepting these messages, phone and radio silence was observed. In addition, cell phone coverage in this part of Texas can be a sometimes thing."

"Does it follow then, that once deployed and acting under your guidelines and their oaths, those involved were able to act independently without seeking your approval for actions that needed to be taken on the spur of the moment?"

"Consistent with the laws of Texas and their oaths, members of the posse were allowed a degree of autonomy in their actions."

"Would this autonomy include such acts as the brave shooting one of the gang members in the shoulder with an arrow in self-defense?"

"It would."

"This complex operation was planned during approximately a 48-hour period during the afternoon, evening, and next day. Is that correct?"

"It is."

"If you had more time, would you have ordered a helicopter crew to assist you in that operation in case the kidnappers attempted to escape by air?"

"Objection, Your Honor," Brown interrupted. "This calls for the witness to make a supposition about actions that did not occur during the event in question."

"Sustained."

"Sheriff Jones. Why didn't you order a helicopter to assist in the operation?"

"This would have taken days to locate a suitable aircraft, obtain approval for it and fly it to the site. By the time a helicopter arrived with a police or military team on it, the operation would have been over. In addition, the costs to the county could have run in the tens or hundreds of thousands of dollars."

"Assuming that you had been able to get such a helicopter, would they be operating under the same rules as the other members of the posse?"

"Objection Your Honor," Brown interrupted.

"I will allow this question," Judge Prescott ruled.

"They would."

"Sheriff Jones, if one of your officers is chasing a criminal who is attempting to escape the scene of a crime in a car would the officers be allowed to shoot at the vehicle in an attempt to disable it and subsequently capture the criminals?"

"They would."

"If a police aircraft was chasing a criminal attempting to escape in another aircraft would they be allowed to fire on the aircraft in an attempt to bring it to earth and capture the criminals?"

"Objection, Your Honor. "This line of questioning is pointless as no police aircraft were involved."

"Your Honor I submit that the Austerhouse Cessna once he was deputized and the aircraft was used to drop the ransom money that it became a police aircraft flown and manned by duly authorized and empowered deputies with the same rights and responsibilities of other officers of the law."

"Objection overruled. Sheriff Jones you may answer the question."

"I am not an expert on aviation law, but my opinion is that such officers would be allowed to fire on an escaping aircraft in an attempt to disable it."

"No further questions," Your witness Mr. Brown.

"Sheriff Jones. Law enforcement officers go through extensive education programs before they are allowed to join the force are they not?" Brown asked.

"Yes they are."

"Many of your officers have degrees in some aspects of criminal justice from various universities and academies do they not?"

"They do," Sheriff Jones affirmed.

"These officers have many hours of training specifically on the legalities of when they may use their firearms and other law-enforcement tools do they not?"

"Yes. They do."

"Larry Austerhouse Senior is a business man and holds business degrees while Albert Hopkins is a pre-med student. Do you have any indication that they have received any training in any aspect of law-enforcement?"

"I do not know."

"Do they have any other qualifications that you know of to guide either of them in the use of deadly force?"

"Objection Your Honor. Without access to exhaustive academic records the Sheriff could not possibly answer that question."

"I withdraw the question. No further questions, Your Honor."

"Is the Defense ready to proceed?" Judge Prescott asked.

"We are Your Honor," Richards replied. "I call Leo Nostrum, County Cornier of Brewster County."

Nostrum the local funeral director and Cornier was a tall middle-aged man wearing a brown suite with a rose in his lapel. He purposefully walked up to the witness stand, took his oath and appeared eager to speak.

"Mr. Nostrum how long have you been Cornier for Brewster County?"

"Approximately 10 years," Nostrum replied.

"What was your training?"

"I was brought up in the funeral industry. I received professional training at Texas State and worked at the Coroner's office in Dallas for 10 years before coming here."

"Have you had experience in examining those who died from gunshot wounds?"

"Yes I have. In Dallas we might have several cases a week doing autopsies of people who died of gunshot wounds."

"Did you have occasion to do autopsies of El Jefe Gonzales and Bill Prospero."

"I did."

"On either body did you find any indications of gunshot wounds?"

"Gonzales had a bullet hole in the neck."

"Did you specifically look for such wounds?"

"Yes, I did. I was told that the aircraft had bullet holes in it, so I was particularly careful about searching for possible bullet wounds."

"No further questions. Your witness Mr. Brown.

"Were there extensive wounds on the body of Bill Prospero and even more on Gonzales' burned body?" Brown asked.

"There were relatively minor cuts and burns on Prospero's body, but the fatal wound was from a 10-inch piece of broken plastic from the helicopter's bubble that drove into his chest."

"You are certain that there were no .30-caliber puncture wounds through the skin that might be masked by superficial burns or other scars?"

"I am because the bullet would have also caused internal wounds, and none were seen."

"Would you describe the condition of Gonzales's body."

"Gonzales had a simple fracture of the left leg, but portions of his intestines, stomach, and most of his liver and spleen were missing were blown away during the grenade explosion during the second gunfight. He also had a single bullet strike on his neck."

"Were you able to determine the cause of death?"

"Gonzales died as a result of lung damage while being burned to death."

"Would the bullet wound in the neck been sufficient to kill him?" Brown questioned.

"The amount of blood being pumped into the body cavity indicated that he was alive during the fire until his lungs stopped functioning."

"Why wasn't he able to escape?"

"His legs were badly damaged by the grenade, and the shock to the spine by the bullet likely immobilized him."

"Have you seen bodies that were this badly damaged before?"

"Yes. I am sometimes called in to investigate burn victims to determine if they were shot before the bodies burned. I can usually make a determination."

"With this massive amount of internal damage and missing organs, is it possible for you to state absolutely that no other bullets passed through the body at some time prior to the body being burned."

"I could find no evidence of such a bullet, but I cannot absolutely say that one did not penetrate the missing organs."

"No further questions," Brown concluded.

"You may step down, Mr. Nostrum," Judge Prescott instructed. "We will break for lunch and when we return Mr. Robert you may call your next witness."

"The Defense calls Albert Hopkins."

Albert somewhat hesitantly approached the stand with his mind racing back to his days in the orphanage when he was whipped for any minor infractions of the many rules that were sometimes enforced with sadistic enthusiasm by the teachers and administrators.

"Mr. Hopkins you were a witness or participant in many parts of the events that have been described in this courtroom, were you not?"

"Yes. I was."

"Did you receive instruction on the use of the AR-15 rifle on the evidence table from Larry Austerhouse Jr. at the ranch?"

"Yes, I did."

"What was the purpose of that instruction?"

"It was to teach me how to use that rifle to hunt deer and possibly defend the ranch if the need ever arose."

"How did you feel about the rifle?"

"It was big, noisy, heavy and it kicked; but I was pleased that I could shoot it. Towards the end I was shooting some groups that Larry Jr. and I were both pleased with."

"Did you find yourself wanting to own a gun like that?"

"Not really. Since Larry Jr. was the only person I knew who hunted and had land to hunt on, I thought I would use one of his guns when I came down to hunt."

"As part of Larry Austerhouse Senior's rehabilitation, you went flying with him on three occasions. Is that correct?"

"It is."

"What did he teach you to do?"

"He taught me how to keep the aircraft straight and level and keep it on a given heading."

"How did you communicate in the aircraft?"

"We had headsets and an internal radio. Mr. Austerhouse could contact the tower at the Alpine airport, and on our second flight we went there to pick up the ransom money."

"At the time you made the first flight to The Solitario you felt confident that you could act as co-pilot?"

"I did."

"Moving to the meeting before the rescue, did you take the oath and receive the signed card from the Sheriff when he swore in the members of the posee?"

"I did."

"What did you do with it?"

"I signed the card and put it in my billfold like everyone else."

"Did you have that card with you throughout the next day's activities?"

"I did."

"At what time did you learn of Austerhouse Senior's plan to potentially engage any helicopter that Gonzales might use to escape."

"That was before we left the ranch to go to the airfield for the first flight with the ransom money."

"Did Mr. Austerhouse tell you what to do?"

"He said that if Gonzales was trying to escape with in a helicopter, I was to shoot at the tail rotor of the aircraft, the tail and the engine; but not at the bubble where the pilot and Gonzales were sitting. The purpose was to bring down the helicopter, and not kill them."

"Why did you agree to do this?"

"Larry was my best friend, and I wanted to do anything I could to help arrest the kidnappers who took him. At the time I did not know that they had also badly hurt him. I don't know quite how to put this, but I also felt strongly bonded to the Austerhouse family at this stage, and would do anything to help them recover their son. They had become family to me."

"When you saw the helicopter and went after it, what happened?"

"On the first pass we surprised the helicopter, flew alongside it and I managed to get off four shots at the machine, but apparently did not hit anything vital. When we came around for a second attempt the helicopter had speeded up and Mr. Austerhouse slowed the Cessna down so I had more time to shoot. A man in the chopper shot at us and hit the Cessna's wing. We pulled off and I tried for him at longer ranges. Ultimately the engine was smoking and we got closer and I got some rounds into the mechanics of the machine and brought it down with my last eight shots. One hit somewhere in the tail rotor and the helicopter started to spin. Another apparently struck the engine and it started to smoke. The helicopter when down spinning."

"You later flew over the crash?"

"Yes we did. We spotted the smoke and found it. There was a person lying beside the aircraft who was not moving."

"Did Mr. Austerhouse take any action at that time?"

"Yes. He radioed back to the ranch mechanic for him to notify the tower in Alpine that a helicopter had crashed and was smoking at a location south of The Solitario. He also instructed Samantha to call Sheriff Jones to notify him about the downed chopper so he could send someone to the site before dark."

"How did you feel?"

"I was disappointed that although we had tried, we had apparently not got Gonzales. I was hoping that the posse could get there before he got away."

"Now moving to the night of the fiesta when the ranch was attacked by the *Bastidos'* Gang. What part did you play in that?"

"The security team's cameras informed us that they were coming. While the cars were being moved to the ranch gate, I was helping Aron and Larry Jr., who were both on crutches into a room with Samantha and her baby. When they were settled, I went out and started walking towards the gate where the other men were gathered.

"While they were talking and giving the beer out, I moved to a slightly elevated position where I could see both sides of the gate. In the light I could see Samantha carrying Michael with a large man behind her with his hands acting as if he were holding her. I got closer and saw that he was holding a knife at her throat. She dropped something bent over to get it and broke away. Sheriff Jones stepped between them. I had a clear shot at his head for an instant and took it at a range of about 30 yards. I killed the man, but not before he sliced Sheriff Jones across the stomach.

"A general round of shooting back and forth began, and I fired a number of shots into the motorcycle gang aiming at their bikes and anyone who was shooting. I emptied the 20-round magazine and stopped shooting. After everything was over I went down to help the wounded.

"It was then that Larry Sr. approached the gate using his walker and he had a pistol in his hand. Although Gonzales was slumped over in the side car, his hands were not visible. One of the ranch hands, Slim Cooper, threw a lasso around Gonzales. When he did Gonzales raised up to throw a grenade. I rushed over knocked Larry Sr.'s walker away and fell on him before Gonzales dropped the grenade into the side car and it exploded. When I hit the walker, the pistol fired and a round hit the top of the side car and passed into Gonzales' neck, as I later found out. That bullet grazed me on the side, but caused no real damage."

"Do you know if you hurt or killed any of the gang members?"

"Perhaps one or two. I don't really know. They started holding up their hands, and within a couple of minutes the shooting was all over."

"How did you feel about the experience?"

"It took me about five minutes to recover from the shock of the blast. Later I felt excited, scared, sad, happy all at once. I was more concerned with keeping people from bleeding to death than anything else. I didn't know if anyone on our side was hurt, but some of the gang certainly were."

"No more questions. Your witness Mr. Brown.

"Mr. Hopkins your actions resulted in the deaths of one person in the helicopter crash and at least one more during the later fight and perhaps wounding or killing two or three others. Would you do so again under the same circumstances?" Brown Began.

"Objection, Your Honor," Richards interjected. "Counsel is making an obvious attempt to paint my client as a habitual criminal."

"Your Honor I only wish to demonstrate the obvious fact that repetition breeds familiarity. That a person who does an act multiple times is apt to repeat that behavior."

"Objection sustained. Proceed with another line of questioning."

"After the fight at the gate you were seen standing over a wounded man and urinating on him. Can you explain this action?"

"The man had a wound on his torso that was bleeding heavily and dirt encrusted. I peed on him as an expedient way to quickly clean the wound before he could apply a piece of torn-up clothing to help stop the bleeding. When we got him back to the ranch, the wound was cleaned more thoroughly, antiseptics applied and so far as I know he did fine afterwards."

"As a relatively inexperienced rifleman you made an excellent shot that killed a gang member who had a knife at Samantha Goldfarb's throat. In front of her she was holding her baby Michael and Sheriff Jones was standing an arm's length away. "Why did you think that you could make such a shot without killing one of these other people?"

"I had a clear shot, the target was exposed, and I acted instinctively. I can't say I really thought about it. I just did it."

"How without specific combat training in military or police units did you come by this instinct?"

"I don't know."

"Don't you realize that you were engaging in an act of 'reckless endangerment' that only blind luck intervened to direct that bullet into a criminal's head instead of an innocent woman, a child, or the Sheriff?"

"Objection. Council is again trying to enter prejudicial statements into the trial."

"Your Honor. I would like to answer that question if I may," Hopkins responded.

"If the Defense has no objection, proceed."

Richards looked carefully at Hopkins on the stand, and replied, "Go ahead then. Answer the question."

"With that rifle and that ammo at a range of 30-yard I was, and am, confident that I could put a bullet into the man's brain at that distance even with the limited amount of shooting I had done with the rifle."

"Would you be willing to demonstrate that to this court," Brown asked.

"The court does not have time for 'play-acting', Judge Pederson added. "I will allow two hours for a demonstration if both parties agree."

Albert nodded his head vigorously and Richards rose to speak. "Your honor the Defense agrees to the demonstration. As Mr. Neuville has been designated as an Expert Witness on firearm matters by both parties, I suggest that he set up the demonstration while the trial continues in court and notify the court when he is ready. Is this agreeable to the Prosecution?"

"It is Your Honor," Richards responded.

"Very well, Mr. Neuville you may remove the gun from the court and set up somewhere safe to have the demonstration with enough room to allow for the jury and court functionaries to witness the results," Judge Presscott ordered.

"Mr. Brown, you may continue questioning the witness."

Brown rose and continued his cross examination.

"At what stage did Mr. Austerhouse Senior direct you to kill the people in the helicopter."

"He never did."

"Did you not consider that shooting down a flying helicopter would likely cause the death of those inside?"

"No. Not really. I was only concerned with getting the rifle ready and in position to make the shots as we flew by."

"In Naval terms a vessel about to be engaged by another is giving a warning 'shot across the bow' to encourage it to surrender. Was any such shot give to the helicopter?"

"No. It was not."

"Was any attempt made to give the helicopter any chance of

surrendering such as contacting it by radio or shooting a flare before shooting at it?"

"No it was not. We did not think of it."

"Do you think that you could have and should have given such a warning?"

"In hindsight I guess we should have."

"No further questions."

"Your Honor, my final witness tomorrow will be Mr. Austerhouse Sr. I expect that his testimony will allow time for both parties to make final statements and conclude the trial. If it please the court, I suggest that the trial resume tomorrow morning.

Neuville walked into the courtroom and went to the Defense table to speak to Richards.

"Your Honor," Richards began, "Mr. Neuville and Sheriff Jones have set up the demonstration and has buses outside to take the court to the site. May I suggest that we hold the demonstration now and resume the trial in the morning.

"Rather than interrupt tomorrow's proceedings, I will allow the demonstration to proceed. The audience will remain seated until the court has departed."

In a pasture outside of town a round Styrofoam head had been placed on a piece of lath while 30-yards away a table was set up with the AR and a box of ammunition.

"Are the conditions of the test suitable for both parties?" Judge Presscott asked.

After obtaining agreement from both sides Albert walked up to the table. Neuville loaded one round in the gun and put the safety on.

"Raise the rifle to your shoulder and shoot as rapidly as possible," Neuville instructed.

Albert felt a flutter in his stomach. He knew he could make the shot, but he had to make a good impression on the jury that he could do what he said he could do. His, Larry Sr.'s and maybe Samantha's life depended on it. He took the safety off the gun raised it to his shoulder, aimed at the head and fired. The head quivered and tiny particles flew out of the back of the manikin unseen by most of the observers.

"You missed," Brown intoned.

"I don't think so," Albert replied.

"Bring the target forward so it can be examined by the court," Prescott ordered.

The Styrofoam head was examined by the judge and then passed between the jury members.

"Let the record show that there is a bullet hole through the head, and that this head be entered as a piece of evidence in this trial."

"Your Honor, this shot demonstrated that the defendant can hit a stationary target at 30-yards. The shot at the ranch was made at a moving target. I suggest that the defendant demonstrate that he can make a more difficult shot."

"Objection your honor," Brown interjected. "The Prosecution asked for a demonstration, and it has been held. In fact my client did hit the knife-holding bandit's head at 30 yards and possibly saved the lives of Sheriff Jones, Samantha Goldfarb and the child. No further demonstrations are warranted."

"This demonstration is concluded, and the Court is dismissed. The trial will resume at 9:00 A.M. tomorrow morning.

Chapter 4

Trial Day 4

"I don't know what the outcome will be, Larry Senior stated with a hint of weariness in his voice; "but either way I will be glad to get this trial over with."

"Remember, that you can take the 5th on these questions if you feel like they would be self-incriminating, Richards reminded the elder Austerhouse."

"In that case I would look like a fool to sit up there like a stump and not say anything. No. I'll testify all right. They need to know what kind of SOB that damn Gonzales was."

"Remember," Richards responded. "Testimony will only be permitted about the facts that you knew at the time – not what you might have learned about later."

While the previous days of the trial had been sunny, the morning brought black scudding clouds whipping across the green grasslands around Alpine that threatened to bring driving rain. Glad to beat the coming storm, those attending the trial entered the courtroom. Several were carrying umbrellas which were examined by the guards.

Last to enter the courtroom were the 10 members of the press who were sat on a reserved space on the third row of seats behind the family members. Judge Prescott quickly opened the proceedings. "The Defense may call its witness."

"I call Larry Austerhouse Senior to the stand," Richards began.

"Austerhouse unfolded his walker and moved from an outside seat down the aisle and Robert opened the gate to allow him to walk towards the witness chair and seat himself. After Larry Sr. sat, he refolded the walker and sat it beside the chair and received his oath.

"Mr. Austerhouse are you the father of Larry Austerhouse Jr., the owner of the Austerhouse ranch and of the single-engine Cessna aircraft that was mentioned in previous testimony?"

"I am."

"What was your relation to Larry Austerhouse Jr., Aron Goldfarb and Albert Hopkins?"

"Larry Austerhouse Jr. is my son, Aron Goldfarb is the husband of my daughter-in-law's sister and Albert Hopkins is a friend of my son who was assisting me and my wife get over COVID-19."

"Before the events that later took place at The Solitario did you receive any communications concerning your son and Aron Goldfarb?"

"Yes. I received a telephone call from a person I now know to be El Jefe Gonzales informing me that they had Larry Jr. and Aron, and were demanding $400,000 dollars ransom for them."

"Did you receive a subsequent telephone call from this individual?"

"I did. During this call the arrangements for the delivery of the ransom money were made, and he also told me that I would receive a present in the mail to indicate that he was serious."

"And what was that present."

"It was a freshly severed human toe."

"Who's toe was it?"

"I did not know. Sheriff Jones examined it with a magnifying glass and said he thought that it was from a Hispanic man."

"Did this convince you that Larry Jr. and Aron were in danger of being killed if the ransom demands were not met?"

"It did."

"Did others at the meeting confirm that Gonzales had a bad reputation?"

"Yes. Both Whitefeather and Sheriff Jones said as much."

"Did you have a real fear that Gonzales was capable of killing your son and Aron Goldfarb or anyone else if he had the chance?"

"I did."

"We have heard how the rescue plan was formulated and the deployment of various groups. Were you a part of this plan?"

"Yes. I and Albert Hopkins were to fly a duffle bag full of the ransom money to a location on the outside of the camp, scatter that money across the desert and then return to the ranch."

"At the meeting on the evening before the rescue or at any other time at the ranch, was there a mention of a second flight?"

"No. There was not."

"When did you decide to take a gun with you in the Cessna and have Albert Hopkins search for a helicopter that might be at The Solitario?"

"When I woke up the morning before we flew, the question came to me as how was Gonzales going to get away? He was at least a two hour drive from the Mexican border. If he had the hostages with him and released them somewhere close to the border, he could be easily captured at the regular border crossing or wherever he tried to cross.

"I decided that he might already have a helicopter ready for his escape somewhere within The Solitario or he might call one in from Mexico to get him and the ransom out."

"You mean you thought he would abandon his men, take the ransom and escape by himself?"

"Yes. That is exactly what I thought."

"Did you discuss this possibility with anyone before you left the ranch?"

"No. When Albert and I left to go to the hanger I had him get the AR with a loaded extended magazine and put it in the truck before driving to the airfield."

"He did not ask you what that gun was for?"

"No. I commonly carry a rifle in the truck with me when I am out on the ranch."

"How far is your hanger from the ranch house?"

"About four miles."

"During that time what did you talk about?"

"We discussed how to tie the duffle back onto the wing struts. It had to be tied so that it would stay there during the flight and then release it so it would fall with a parachute that would open the bag's zipper and spill the money on the desert floor outside of the camp. Everything had to be tied on so that no cords would be tangled in the landing gear."

"This sounds like it might be complicated. Was it tested before the drop was made?"

"We threw it off a windmill, and it worked after I put a bag of pesos in it to make it fall straight. We still had to figure out how to tie in onto the aircraft, and my mechanic helped us do that when we got to the strip.

"Albert had to pull two ropes - the first one to cause the bag to hang straight down and the second one to release it and allow the parachute to deploy and scatter the money."

"The drop was successful?"

"Yes. It was."

"What happened after you returned to the strip?"

"We refilled the Cessna, and I told Alfred to get the AR, put the magazine in it and get inside the aircraft. I then told him that we were going to look for a helicopter in The Solitario and attempt to disable it if we found one. I told him to shoot at the tail rotor and engine and not into the bubble. Our objective was to disable the helicopter and not kill the people inside – just prevent them from getting away. I also asked him if he thought he could do this, and he said that he could."

"How much time was it from the time you made the money drop until you were back over The Solitario again?"

"I didn't really look – perhaps an hour or a bit longer."

"When you arrived what did you see?"

"My hands and the caballeros were rounding up the gang. I also saw Larry Jr. and Aron and they waved that they were all right."

"What did you do then?"

"I circled The Solitario to gain some altitude and that is when we saw a small Bell helicopter taking off from a feeding area near a water tank."

"What direction was it headed?"

"It was headed south towards the Mexican border."

"What did you do then?"

"I circled over it so that I could come alongside and told Albert to get ready to shoot when I made a pass alongside the helicopter."

"On that first pass was Albert able to shoot?"

"He shot several times – four I think."

"What happened on your second attempt?"

Using his hands Larry Sr. demonstrated as he spoke. "The helicopter was like my right hand here – below the Cessna. It had speeded up and I slowed my airspeed so that I more-or-less matched it and flew a parallel course passing above it and some 20 to 30 yards from the side of the chopper. The helicopter whipped around, and a man fired a shot which hit the Cessna. We made several other passes and shot at longer range until we saw smoke coming from the engine. I got closer, and Albert started shooting. In that last attempt, I think he fired seven or eight rounds before I pulled up away."

"Once you pulled away could you see the helicopter?"

"I could not, but Albert said that he hit it and that it was going down."

"Did you take any further action?"

"We went back to that approximate area and looked for the helicopter. Albert spotted some smoke and when we flew over that he said that the chopper was down, and he could see a man on the ground beside it. I over flew the crash site again and saw the same thing."

"What were your next actions?"

"I started back to the ranch and radioed to the ranch that Larry Jr. and Aron were all right and that there was a crashed, smoking helicopter on the ground south of The Solitario north of the South Fork Road. I asked that this information be telephoned to Sheriff Jones and the tower at Alpine."

"Did you tell Sheriff Jones that you and Albert had shot down the helicopter?"

"No. I did not."

"Did you tell anyone that day that you had shot down the helicopter?"

"I told Mrs. Goldfarb when I got back to the ranch."

"Didn't you think that anyone else would want to know that information?"

"I figured that if Sheriff Jones or anyone else wanted to know about it that they would ask."

"Did they ask?"

"Yes, they did during the accident investigation, and I told them exactly what happened."

"When questioned, you did not attempt to conceal any of the facts of the case?"

"No. I did not."

"Did you feel like you were acting as a private citizen seeking vengeance or as a sworn law enforcement officer doing your best to apprehend a violent and desperate criminal?"

"Objection Your Honor. The accused's feelings in the matter are not pertinent to a determination of his guilt or innocence of these charges."

"If the court please, the Prosecution has stated that he is going to attempt to prove motivation for this alleged crime. In this even the Defense is entitled to examine questions regarding the intent of the accused."

"Objection overruled. The witness may answer the question."

"I admit that I did want to see this criminal brought to justice because of the harm that he had inflicted on my family, but I felt that my stopping him from escaping was a legal duty, rather than a matter of personal vengeance."

"No more questions. Your witness Mr. Brown."

"Mr. Austerhouse. Do you consider yourself a law-abiding citizen?"

"Yes. I do."

"Are you aware that it is necessary to be licensed to give flying instructions in the state of Texas?"

"I am."

"Would you call your teaching Albert Hopkins how to fly giving flying instructions?"

"Albert, acting as a Physical Therapist, was helping me regain my strength and abilities from a severe case of COVID-19. Controlling an aircraft takes coordinated skills that tired me out. He was there as a back-up in case I did not have the strength to control the Cessna."

"Once in the air, is the Cessna a difficult aircraft to fly."

"No. It is not. The tricky part is in take-offs and landings."

"On your flying trip to Alpine for the ransom money, he was largely in control of the aircraft for how long?"

"Approximately two hours."

"While I can see the possibility of a pilot giving control of an aircraft to a passenger for a few seconds so that he might feel something of the flying experience, a two-hour trip would seem to be more like a lesson than a thrill ride would it not?"

"That is how I was taught and all pilots once were taught."

"That was not the question. Did you in fact give unauthorized flying lessons to Albert Hopkins in violation of the law?"

"Objection Your Honor. The Counsel is again attempting to get the witness to incriminate himself."

"Overruled. The witness will answer the question."

"If you put the question like that, I guess so. A two-hour trip would fall more into the lesson category than a flying demonstration."

"You own to AR-15 –style rifles and at least one extended magazine that was under a national band?"

"I do, but these were legally purchased after the band was lifted."

"Why do you own these guns that were once legally disapproved by society?"

"I wanted them for hunting uses and to defend the ranch if necessary."

"Mr. Austerhouse. You have in your employ a number of ranch hands some of which you have employed for many years and others of which are more recently employed. Are any of these illegal aliens?"

"Not at this time. Some are temporary workers with green cards, some so-called dreamers and others are U.S. residents."

"You say 'not at this time.' Does that mean that your ranch once employed illegal aliens?"

"Before Nixon's general amnesty law there were some, but they are now U.S. citizens."

"On your ranch you have a large number of vehicles and trailers do you not?

"Yes. I do."

"Are all of these legally licensed under Texas laws?"

"We usually don't put plates on vehicles that we operate mostly on ranch lands that do not get out on the public highway. These are mostly trailers for hauling stock, tractors and some of the older pick-ups."

"These never cross public roads?"

"Sometimes they do."

"They never go on public roads traveling from one part of the ranch to another?"

"They sometimes do travel these roads for short distances."

"Objection Your Honor. I fail to see how this line of questioning has any significance."

"Sustained. Please move on, Mr. Brown."

"You were in radio contact with the ranch and through telephone relays to Sheriff Jones. Could you have gotten a message to Sheriff Jones and had him send a helicopter to cut off Gonzales' helicopter before it reached the border?"

"I did not think that there was time for anything like that to happen since the nearest such helicopters that I knew anything about would have had to have come from the army or National Guard in El Paso. It would likely have taken days to get something like this arranged."

"But you did not try, and now we will never know. Will we?"

"No. We will not."

"You also never attempted to contact Gonzales by telephone, radio or by signal from your aircraft for him to put down and surrender before or after you started shooting at him?"

"No. I did not."

"Why?"

"I knew that he had already cut off a man's toe. I expected that he was armed and just as I might be able to shoot him down, so could he. I did not want to give him a chance to shoot at us."

"So you snuck up on him and ambushed him without warning."

"I did."

"You had a second opportunity to shoot Gonzales at the ranch gate, and in fact did shoot him in the neck. Is that correct?"

"It is. I thought he was playing opossum."

"You mean that he was not responding to your demands to raise his hands, and so you shot him?"

"I did not mean to shoot him unless he did something dangerous. Slim stopped me from getting closer and threw a lasso around him. When he did Gonzales sat up and attempted to throw a grenade. Slim tightened the noose causing him to drop the grenade into the side car. Then Albert knocked me down and fell on top of me. When he did, the gun went off, and the bullet apparently hit Gonzales. I did not purposefully shoot him, although I would have under the circumstances."

"Your hands were on the gun, and you pulled the trigger. You shot the man you hated, didn't you?"

"Yes. I did. I did not mean to at the time, but I did."

"I submit that you did not manage to kill him on your first attempt, so you were going to make doubly sure that you were going to kill him now that you had a second opportunity?"

"Objection, Your Honor." Richard protested. "This is conjecture on part of the Prosecution."

"Sustained," Judge Prescott ruled.

"No further questions," Brown concluded. "Your witness, Mr. Richards."

"Does the Defense have any questions on redirect?" Judge Prescott asked.

"I do not Your Honor. The Defense rest."

Counsels will present their closing statements after lunch. This count is adjourned until 1:00 PM."

Chapter 5

Judgement

"**M**R. BROWN, IS THE PROSECUTION prepared to present your closing argument?"

"If it please the court, Your Honor. This is a very simple case. As the jury has seen there are few real conflict issues between the Prosecutor's and Defense's presentations. The agreed-on facts are One: There was a planning meeting at the Austerhouse ranch the night before the rescue that was attended by the Defendant and significantly in the presence of Sheriff Jones. Two: Mr. Austerhouse Senior with the assistance of Albert Hopkins was to fly the Cessna aircraft over the kidnapper's camp and drop the money to distract the kidnappers. Three: This plan was approved by all, and Mrs. Samantha Goldfarb acted as Director or Coordinator.

"Four: Members of the rescue teams were sworn in by the Sheriff as legal members of a Posse. Five: No mention or authorization of a second flight by the Cessna was approved or addressed. Six: A second flight known only to Austerhouse Senior, Anthony Hopkins and the Mechanic Blair Adams was made to The Solitario later that morning. Seven: A Daniel Defense .308 rifle was loaded into the aircraft before departure. Eight: Larry Austerhouse Senior told Albert Hopkins to shoot to disable any helicopter that was seen. Nine: No attempt by any method was used to obtain permission from Sheriff Jones to take this action. Ten: No attempt was made to contact the helicopter to encourage it to set down or return to the camp. Eleven: In at least four attempts Albert Hopkins shot into the helicopter forcing it down and ultimately killing Bill Prospero and the destruction of the helicopter. Twelve: in the later fight at the ranch Larry Sr. shot Gonzales which resulted in his being unable to escape the fire which burned him to death.

"These are the uncontested facts of the case as presented by the witnesses for both the Prosecution and Defense.

"The Defendant, Larry Austerhouse Senior is charged with Capital Murder, the most severe charge in Texas statutes. This charge carries with it the duty that we must prove beyond a reasonable doubt that with planning and forethought the defendant did plan and cause the deaths of these two individuals, although it took two attempts to kill Gonzales. It is not necessary for us to have proven that he 'pulled the trigger' in that he killed these individuals by himself. What is necessary is that we have proved that he planned the events that led to the deaths of Pepe Gonzales and Bill Prospero.

"We have the defendant's testimony that he ordered Albert Hopkins to load the Daniel Defense AR into the aircraft and instructed him how to shoot down the aircraft. This testimony is supported by Hopkins' own descriptions of the events.

"We have further testimony that the death of Bill Prospero was directly caused by the crash of the helicopter. We also have testimony from the Coroner that Gonzales was injured with a broken leg during the crash and because of his injury and a subsequent bullet wound was unable to escape being burned alive.

"These deaths would not have occurred had Prospero and Gonzales been given the opportunity to surrender to legal authorities, which they were not afforded. These deaths would not have occurred unless Larry Austerhouse who had previously shown a flagrant disregard for Texas law had not unlawfully taken the law into his own hands and caused the helicopter to be shot down. His actions directly resulted in the deaths of these two individuals.

"Because there was in this case premeditation and planning and no effort to either obtain permission to execute the chase and shoot down the helicopter or warn the individuals inside the aircraft, the Prosecution calls for the full weight of the laws of Texas to fall upon Larry Austerhouse Sr. for the charge of Capital Murder and will seek the death penalty on this charge."

"Thank you Mr. Brown for your succinct summation of the case for the Prosecution," Judge Prescott commented. "Mr. Richards are you ready to present for the Defense?"

"We are Your Honor. The Prosecution has attempted to paint the Defendant and Albert Hopkins as bloodthirsty vengeful killers who carried out these acts with malice and forethought with a complete disregard of the law.

"We have shown that the rescue operation was successfully executed with the significant aid of Mr. Austerhouse and Albert Hopkins who both contributed to the planning for the rescue of Larry Jr. and Aron Goldfarb without loss of life except for the leader of the gang, El Jefe Gonzales, and the helicopter pilot Bill Prospero.

"We have shown through the testimony of Mr. Austerhouse and Mr. Hopkins that Hopkins was expressly told not to shoot into the bubble of the aircraft or at the people inside, but to aim at the helicopter's mechanisms with the intention to bring the aircraft to earth and not kill the individuals.

"The accident investigator testified under oath that although bullets did strike the body of the aircraft no hits were found in the aircraft's bubble or in the seats of the aircraft to indicate that any bullets struck those inside. The Cornier also testified that he could find no evidence that either individual was hit by bullets fired by Hopkins.

"Neither the helicopter nor the Cessna were designed as combat aircraft and do not carry any armored components that would provide protection for the pilot or passengers. A single shot from an armed desperate fugitive could have downed the Cessna which is why no advance warning was given before the helicopter was shot down.

"The fact that Gonzales had already killed and cut the toe off one of his captives was sufficient to cause Sheriff Jones to call off the Posse from pursuing Gonzales that night because of the strong possibility that he would ambush a member of the posse as they approach at very close range and lives would be needlessly lost. Sheriff Jones prudently called in a dog and trained rescue team to go in after daylight the next day to find Gonzales, but he had escaped only to return at the Fiesta with another gang. This was a man who was recognized before and after the helicopter was shot down as an individual who could be expected to kill anyone who attempted to keep him from escaping.

"Sheriff Jones testified that officers of the law were authorized to shoot at an escaping criminal's vehicle to disable it. We have established that as sworn members of the posse the Defendant and Hopkins were officers of the law at the time they engaged the helicopter, and the resulting deaths of those inside were unfortunate accidental occurrences and not purposefully caused by the individual. The intent of Larry Austerhouse Jr., Albert Hopkins and Mrs. Aron Goldfarb was not to extract vengeance on those who died, but to successfully recover their family members.

"The Prosecution has attempted to paint Albert Hopkins as a bloodthirsty killer who like a lion who has tasted human blood, wants more and more of it to satisfy some evil instinct. This is completely untrue. As a medical student Hopkins fought to save lives in the emergency room and saw firsthand the results of gunshot wounds on the human body. He did, through an exercise of discipline and skill fire multiple shots from a moving aircraft into the operating parts of another moving aircraft without injuring those inside with his bullets. Later he made a close range shot with the same rifle at one of the *Bastidos* gang who was holding Mrs. Goldfarb and her child Michael at knife point. Subsequently he shot at the motorbikes and those gang members who were shooting at fellow family members and others of the fiesta guests who were attempting to repel the attack. After probably saving the life of Larry Sr. by protecting him from a grenade explosion, he administered first-aid to those individuals, which included cleaning some of their wounds with his own urine to help prevent follow-up infections. His actions helped save lives. This was not the act of the bloodthirsty kill-crazy individual that the Prosecution would have you believe that he became.

"The Prosecution would have you believe that Mr. Austerhouse is a brutal land-baron from the 1800s who lives on a vast track of nearly barren land and often has a complete disregard for the law by supposedly giving flying lessons, operating untagged vehicles on public roads, employing illegal aliens and possessing illegal weapons. None of these allegations have any bearing on this case. Yes. Mr. Austerhouse did teach Albert Hopkins how to control the Cessna as a part of his rehabilitation therapy. Yes. Untagged farm vehicles might run on public roads for short distances when going from one part of the ranch to another. Illegal aliens were once employed at the ranch, but this is not the practice today. The AR-style rifles owned by Austerhouse were legally purchased and used for a legal purpose, that is defending his own home when it was attacked.

"The Prosecution has made much of Austerhouse's failure to contact the Sheriff to obtain permission to attack the helicopter. Contrary to what is commonly seen on Television shows, in West Texas telephone and radio communication may be a sometimes thing. Because of the use of different frequencies it is impossible to know without advance knowledge if the units are compatible and can receive each other's signals. Hence the complicated system used to relay the message to the Sheriff. Austerhouse had to first send a radio message to his ranch and then Mrs. Goldfarb had to use a cell phone to call the Sheriff. As only the radio receiver was at the ranch, she would have had to have located the mechanic, driven four miles to the strip, and had him contact the Sheriff. These actions would have had to have been successfully completed before the helicopter crossed the Mexican border. While theoretically possible, this was not doable as a matter of practicality.

"The question before the jury is did Larry Austerhouse Senior act as any prudent West Texas ranch-owner might do to protect his family and home to demonstrate to Gonzales and other potential evildoers 'not to mess with Texas.' The Defense has shown without any question, that he acted within the legal authority granted him as a member of an authorized posse to engage the helicopter to keep a dangerous criminal from escaping. Two men died as an unintended consequence of this action which he attempted to prevent by having Albert Hopkins shoot only at the operating mechanisms of the helicopter and the unintended discharge of his pistol during the later fight.

"We ask the jury to consider all of these points and return a verdict of not guilty of all charges.

"This completes the case for the Defense, Your Honor."

"This court will adjourn for lunch until 1:00 PM. The jury is admonished not to discuss the case or deliberate in any manner until you return. Everyone will remain seated until the jury retires from the court."

"Now we mostly wait," Richards responded once the family was back at the Holland Hotel's dining room. When we go back in the judge will charge the jury and discuss possible verdicts. With anything other than a non-guilty verdict there will be a separate sentencing hearing."

"How long will they be out?" Samantha asked.

"One never knows," Richards replied. "They could deliberate for anything from about 15 minutes to days. This is a capital case. There must be a unanimous verdict. I think that I have presented a good case. However, I gave up a long time ago trying to predict results from jury trials. I frankly don't know."

"Aron, you heard the case. You are perhaps the most objective observer here," Richards commented. "What do you think?"

"If I were in that jury box, I think that the murder charge would be too strong. Whether they might consider a lesser charge, like manslaughter, I don't know."

"Neither does anyone else," Larry Sr. said. "I was watching the jury. They were paying attention to both arguments. They are mentally deliberating them now, despite what the judge said. Whatever they decide, I don't want any appeals. I want to end this now."

"I won't hold you to that," Richards said. "In capital cases there are automatic appeals to allow the case to be independently evaluated. We'll just have to see. Whether you want to or not, all of you eat something. We may be in for a long evening."

"Members of the Jury," Judge Prescott began when the trial resumed. "The state of Texas considers taking a person's life a very serious business. You must reach a unanimous verdict to return a judgment. You may find the Defendant guilty as charged of Capital Murder, or on lesser charges of Murder, Manslaughter or Criminally Negligent Homicide.

"Capital Murder carries a sentence of 99 years imprisonment or the death penalty. This charge is a result of a homicide and multiple other crimes being committed as part of the same offence.

"Murder is a lesser charge which carries a lesser prison term with the possibility of parole. You may find the Defendant guilty of murder if you believe that his action resulted in the death of only one of the two individuals.

"Manslaughter is appropriate in those cases where by a reckless act he caused the death of another person or people.

"Criminally Negligent Homicide may be found in cases where when a death results because of some inaction on the part of the defendant, and is not applicable in this case.

"A finding of Not Guilty means that you the jury did not find that the evidence presented in this trial was sufficient to support any of these charges.

"Yours is the responsibility to find for either the guilt or innocence of the Defendant. The Defendant's sentence will be determined in a separate proceeding where mitigating evidence may be presented. The fact that others who testified may also be facing charges has no bearing on this case. The fact that the individual had a well-known criminal background also has no bearing on your deliberations. Even criminals have the right to defend themselves in court. It should be kept in mind that the helicopter pilot Bill Prospero was not shown to have any part in the gang's criminal activities other than flying a chartered aircraft.

"The jury will now retire and consider its verdict," Judge Prescott concluded." The seven men and five women rose and filed out of the jury box to go to the Jury Room to consider their deliberations. With windows on one side and a giant rendition of the lone-star flag of Texas they settled in to deliberate the case.

Returning to the private dining room of the hotel Bucky Richards remarked. "Whatever the outcome, this may be the first of a long series of trials and appeals as Albert's and Mrs. Goldfarb's cases come to trial."

"I feel relieved that the end of this mess is at least in sight," Larry Austerhouse stated. "I am glad that this thing only lasted four days. I feel better that this stage of this business is over."

Before the group had even finished their coffee, Richards' cell phone rang and a few muddled words were heard. "That was the bailiff. The jury is ready to deliver its verdict."

With a clattering of coffee cups and a scraping of chairs the excited group prepared to return to the courthouse two blocks away. Mr. Austerhouse with his walker was escorted into a waiting vehicle which was to take him Samantha and Michael back to the court, while the others, too impatient to wait walked briskly up the street to what all hoped would be their last visit to the red-brick courthouse which was ominously dripping with water from an earlier shower and seemingly loomed larger over the group when framed by the low grey clouds.

Once the court had been reassembled, Judge Prescott announced, "I have been informed that the jury has reached a verdict. No outburst will be allowed in the court when the verdict is announced under penalty of being charged with Contempt of Court.

"Bailiff please bring the jury into the courtroom."

Everyone examined the faces of the jury as they took their seats, but there was not a hint in their expressions of what the verdict might be.

"Mr. Foreman. Has the jury reached a verdict?" the judge questioned.

A middle-aged man who had the look of a college professor rose and replied, "We have Your Honor."

"Would you please pass your verdict to the Bailiff?" Judge Presscott requested.

The Bailiff took a piece of paper from the Foreman and walked it to the Judge and handed it to him. Adjusting his glasses, Judge Prescott looked over what was written on the paper and once again addressed the foreman. "Would you read the verdict to the court?"

"We the jury find the Defendant Larry Jason Austerhouse Senior not guilty on all charges."

Despite the judge's warning there were gasps and sighs of relief heard in the courtroom, but due to his strong admonishment, these were somewhat subdued as the potential significance of the verdict sunk in.

With a slam of the gavel Judge Prescott demanded, "The court will come to order.

"Mr. Foreman. So say you? So say you all?"

"We do, Your Honor."

"All charges against the Defendant Larry Jason Austerhouse Senior are dismissed.

"Does the Prosecution have any other business before this court?"

"We do Your Honor. It is the opinion of the Prosecution that the state of Texas has spoken in this matter. All charges against Mrs. Samantha Goldfarb and Albert Hopkins are dismissed."

"So be it entered into the record," Judge Prescott ruled. "The jury is discharged with the thanks of the court. This court is adjourned."

With this announcement pandemonium erupted in the courtroom. Hugs and kisses were exchanged and a round of applause erupted from many of the spectators. Justice, although perhaps a bit on the "rough and ready" side, had been served.

Chapter 6

Treatment

"Larry, I know what you told me what it was like up in the kitchen with you, Aron and Henry fighting off those men with flintlock pistols and a sword," Beth began; "but it was no fun in the basement either. Alex was at the top of the stairs with a knife ready to take on anyone who came down. Dad was standing beside the stairs with a grubbing hoe ready to break the legs of anyone who started down the steps while your dad and me were ready to finish off anyone who fell on the floor with a hatchet and spade.

"If anyone had come down those stairs it would have been a bloody mess, and I was ready for it. The thing that was running through my mind was 'How dare they?' I wanted to hack them into little pieces. I really did."

"Beth I can see where you a coming from, but it is where you are going with this anger when school starts that worries me," Larry Jr. replied. "I know the crap that you put up with the School Board about your Special-Ed programs for your kids. Are you going to want to hack those Board members up too? Or violently confront their parents? I am really scared that you are going to have a violent outburst that is going to result in somebody getting hurt and you losing your job.

"Even more important than that, those kids would have to start all over again with someone they did not know."

"That's tearing me up too." Beth continued. "I know they are not my kids, our kids; but I feel after so many years that they are. They tried this remote-learning crap, and God knows I pulled every string I could to make it work, but they lost so much ground on their developmental milestones.

"I fought for those kids. I fought hard to get computers, internet, training and even food in their mouths and keeping a roof over their heads with some of the parents who really didn't know what their children needed or how to provide it. I had to be there. I wanted to be there. I needed to be there; but I could not be there.

"The possibility that I could get in a rage over some minor shit with the Board and get me fired is bothering me too."

"While we were in Texas, I had Albert gather as much information as he could about peyote, which Aron and I ate. He found suggestions that it might be able to moderate long-term anger. I also talked to Whitefeather about their experiences with it in the Native American Church and with Reverent Billy as well as your dad. They said they had positive experiences with it, but knew of others whose visions took them in terrifying directions. On the clinical side there are a few small studies about its use in perhaps handling some psychological problems, including one involving some researchers here in Atlanta.

"Each potential study subject is evaluated for the trial and the drug is administered under controlled conditions. I talked about this with your dad who took it years ago, and he said he thought it was worth a try. Do you want to sign up for the trial?"

"I don't know that I want to tell anyone about this stuff," Beth snapped back. "There is just too much personal shit."

"Individual personal information is kept confidential. The patient is never named in the reports or where they worked or other information other than things that might impact the performance of the drug. No one need to know about this study except you, me, the doctor and anyone you might want to tell like your dad, mom and sisters."

"Sam is better at this cost-risk analysis that I am, but I think that

I am ready to do something if it keeps me from doing shit like I did with Aron last year." Beth smiled as she remembered the event. "That is funny to think about now, but was not at the time. I put poor Aron through hell."

"Mrs. Austerhouse I am glad that both you and your husband could come," Dr. Mayoanne Guptaglong welcomed. "I have been working as a psychologist with hallucinogenic drugs for several years attempting to discover if they can be used to treat various disorders. This particular trial is with doses of mescaline which is the active ingredient in the peyote cactus. This cactus has been used for centuries in Mexico to produce magical visions which those cultures interpreted as being inspired by their gods.

"Any study of this sort controls as many variables as possible. Some, like past experiences are not controllable. Others such as the number and types of doses and the manner in which they are administered we can, and do, control very carefully. These drugs give hallucinations along with physical symptoms, but we have no advance notions about what these might be. The hallucinations have been described as pleasant, unpleasant, but somehow comforting, horrible and everything in between.

"We try to put the patient in a relaxed meditative state in a dimly-lit room with as little outside stimulation as possible. Commonly the experience is somehow derived from something seen in the past. It could be a traumatic experience or a pleasant one. It is like all of your memories were on a record and we are dropping a needle on a groove. It is from a random bit of memory that the hallucination is derived.

"With most patients these memories are vivid and memorable. Often the patients feel as if their journey through whatever they were experiencing is incomplete, and they want to go back. Unfortunately, they most often do not return to that event again, although some keep trying with repeat doses of the drug.

"The mescaline content of the peyote is variable. When it is eaten it can cause some people to throw up. To exactly control the dose we use mescaline pills instead of having you eat the cactus.

"I am going to give you a small oral dose at first. I will evaluate that reaction and after that increase the dosage to the amount that appears to have a therapeutic impact. Each person's reaction is different, and each event is likely to be different in detail. Your husband, me and a nurse will be with you during the event which may last for half-a-day with lingering responses continuing for even longer. In some cases, with LSD there have apparently been lasting benefits to patients having issues of various sorts. I am going to attempt to determine of mescaline can offer similar benefits.

"There are unfortunately no advance indications that can predict any particular outcome. With most psychological conditions, a combination of drugs and therapy give the most reliable results.

"Beth, have you tried meditative techniques or physical exercise to help alleviate your anger symptoms?"

"Not very successfully," Beth replied. "My sister, Sam, Samantha, and I did try meditation with Dad; but I never had the patience for it. I had always rather be doing something else. Exercise helped take my mind off my problems, but that is not something I could do during a meeting. I just had to suck-it-up and keep myself together. When I got home, this pent-up anger broke out with me taking it out on someone else, like my husband."

"Before you start, I would like you and your husband to try meditation again – together if you possibly can. The reason is I want to establish a baseline from which to start. During the drug administration I will be examining your brain waves to see what parts of the brain are activated by the drug as well as to see if there are any apparent abnormalities. We will also do a MRI of the brain to potentially identify any organic factors that might be influencing your reactions."

"You mean I might have some sort of brain tumor?" Beth questioned with a note of concern.

"There are a variety of tumors and brain conditions that can cause various reactions," Dr. Guptaglong replied. "Because this is apparently a life-long condition, this is not too likely, but something I need to check off my list of possible causes."

"You said that you were going to study brain waves?" Larry Jr. asked. "Are you going to have to shave her head for that?"

"I don't deny that shaving the head would make the process easier," the doctor replied. "With micro-technology we can remove relatively small amounts of hair to install the electrodes. We do this particularly for performers whose hair is such a part of their persona that it would add additional psychological trauma for us to cut it off."

"Beth, how do you feel about that," Larry Jr. asked.

"I need to keep my hair because the children would be so distracted by my being suddenly bald, that I couldn't do a thing with them."

"I want you to both take your time. Try some meditation and make your decision. The meditation will not hurt, and it might be sufficient to solve the problem. After you have considered what I have said and read the materials that I have given you, sign the consent forms and mail them to my nurse. Then we can set up a date for the brain scan and pre-trial blood tests and exams."

"What does the experience feel like?" Beth asked.

"Mescaline often has impacts on how the body feels to the individual. Some want to dance, run or climb. Others will feel like all of their nerve endings are activated and every nerve tingles. You may feel hot for a time or really energized. The response varies depending on where you are and how much drug is in your system."

"I felt like I could still feel its impacts the next day," Larry Jr. added. "Is that typical?"

"The most pseudo-scientific information that I have is from a website called 'Psychynaut Wiki' which reputes to gather information from users and consolidates it. I think that the site may contain some valuable information, but without scientific testing and peer-review, double-blind studies, etc. this information is to be used with great caution. They report that a therapeutic dose of the oral drug takes from 45-90 minutes to produce noticeable impacts. The peak experience last for four or five hours, but aftereffects may persist for as long as 36 hours. So yes, it is entirely possible that you were still experiencing some impacts from the drug the next day. Taking the drug increases urination, and about half the dose is passed in about six hours.

"Our question is does it alter the neuropathways of the brain to the extent that it permanently impacts behavioral change, and if so, what changes?"

Compared to the huge MRI machine enveloping her, the diminutive Beth felt like a pencil rattling around in a cigar tube. She was warned to remain absolutely still during the 40-minutes that the images were being recorded despite the pings, blings and hums emitted by the machine. Finally, the process was completed, and Larry Jr. and the nurse helped her out of the device. She could not help but think how difficult it would be for some of "her kids" to go through such a process.

"For the brain activity test," Dr. Guptaglong explained, "I am going to ask you a series of questions like this was a lie detector. At first these will be ordinary like your name, what is your favorite color and so on. Then I will ask you about pleasurable experiences and conclude with a series of probing questions designed to elicit an anger response. What we want to see is what parts of your brain are activated by which questions and gather information about the degree of activation. I am going to restrain your arms and legs, because if you start moving that will confuse the results. Is this all right with you both?"

"I don't want to hurt you or anyone," Beth assented. "Go ahead."

"If it is all right with Beth, it is all right with me," Larry Jr. assented.

The first series of 20 questions passed without incident. Then they became more personal. "Do you enjoy sex?" followed by "Do you enjoy oral sex?" and then "Do you enjoy anal sex?"

With these invasions of personal privacy, Beth was getting more agitated. The triggering question was, "What if your Board told you that it was a waste of County resources, and they were going to send all of your children home and give you a job in the school kitchen."

"You will do no such damn thing!" Beth erupted with a force that startled both Dr. Guptaglong and Aron. She struggled in the chair, but after a couple of minutes regained her composure.

Dr. Guptaglong resumed his questions, but this time returned to less probing subjects designed to return Beth to a more normal state.

"Thank you both very much," Dr. Guptaglong said at the end of the session. The next session we will administer the drug and it will be all day. You still have the option of withdrawing from the study at any stage. Just notify my nurse. Otherwise, we will see you again in three days. Beth, please do not drink any caffeinated drinks or take any alcohol or other drugs the day before the study. The medications that you regularly take can resume twelve hours after treatment. Do you have any questions?"

Beth looked at Larry Jr. and he shook his head. She spoke for both of them. "I started this thing and I want to finish it. See you in three days." With that comment, they walked out.

Once in the elevator, Larry Jr. asked, "Did you feel like you wanted your hatchet?"

"No, but I did feel like I wanted to whomp him on the side of the head when he started asking about our sex lives and suggested throwing my kids out of school and putting me in the kitchen. He found my pressure points all right."

The rooms where the treatment was to take place had padded gray walls and track lights that could be dimmed. A toilet and sink had been attached in one corner behind a screen. The room was also equipped with a sound system that could play a variety of soothing sounds like gentle rain, waves lapping on a beach and bird songs to provoke a natural environment.

There were three softly upholstered chairs and a retractable table that could be pulled out of the wall. On the table was a tray holding a pitcher of water and a sealed plastic bag containing capsules. One was green and the other two were red. The patient's chair was fitted with electrodes which were attached to Beth and monitored in an adjacent room.

Once the nurse, Aron and Beth were seated and the electrodes attached; Dr. Guptaglong handed Beth the green capsule. "This is the low dose starter. In about 20 minutes you will start feeling something. If you think the experience is tolerable, we will give you the stronger dose second capsule. If that is tolerated well, we will follow with the third. When you need to go to the bathroom let the nurse know and she will disconnect you. This will be an eight-hour event. Larry when you go for lunch or I leave, I will call for another nurse so that there are always two people present with her. We can discontinue the process at any time, but once the higher-level capsule is administered it will take six hours for its impact to diminish to the point where we can put you in a regular hospital room for the remainder of the night.

"Do you want to continue?"

Beth looked at Larry and then put the capsule in her mouth and swallowed. In time she started to feel her body warm and the walls started to show geometric shapes which started to take on hints of color. Although she knew that Larry Jr. and the doctor were beside her, she felt as if she was becoming somehow distanced from them.

"She knew when Dr. Guptaglong offered her the second pill and she took it. Sometime later, she did not know exactly how long, things became more intense. Like lights were being turned up. She could smell Larry Jr.'s musty man smell and the turmeric from Dr. Guptaglong's last meal.

The colors were now vivid yellow, green and purple as sharp-edged geometric patterns formed and dissolved. She reached out to touch one of the shapes but her hand vanished among them until she withdrew it. A play, like a dream but more vivid, started to form in front of her, but there were no boundaries on the stage – only the sand on which the figures stood. She was in it, but it was not in her present form. It was an avatar, an Amazon warrior dressed for business combat in black ninja clothing and carrying a sharp-edged pocketbook which she was using to battle, beat, cut and pummel the smaller scantily clothed men who faced her.

Then there were clouds and sea in brilliant blue and green with smaller figures rising out of the sea walking towards her. They were raising their arms as if they wanted to be picked up. This was followed by the return of the geometric symbols and a feeling of prickling throughout her skin. Wherever she had been for however long it was, she was on her way back from that place whether she wanted to be or not.

Now awake, but in a hospital bed, she looked and saw Larry Jr. beside her in a chair. She reached out towards him, but the IV in her arm immediately stopped her. Larry Jr. noticing the movement, got up from his chair and went to his wife.

"How was your trip?" he asked.

"Interesting, I guess I would say," Beth responded. "I think that I got to kill some mental demons and maybe was rewarded for it. It was something like a dream, only more real. It was like comparing a black and white movie with a Technicolor one.

"I feel maybe a bit more at peace with myself and everything. Like maybe I can take on those bastards if I have to and win – win for me and win for my kids."

Chapter 7

Graduation 2

"Are you ready yet?" Samantha shouted up the stairs where Aron was getting Michael ready for his graduation from Lakeside High in just two hours.

Mom, I am about as ready as a guy wearing a dress and the most stupid looking hat I have ever seen can be." Michael replied as he stood at the top of the stairs.

"Hold it. I want a picture of you and your dad looking half-way descent on the stairs." Samantha ordered as she got her camera ready. Although it was a bit tight with them both standing side-by-side on the stairs, Samantha managed to pose an appropriate pre-graduation photo with both father and son looking at camera and smiling. Aron had his arm around his son while Michael was restraining himself to keep from bursting out laughing.

While Aron still had more body mass, Michael matched him in height which had aided him as a forward on the high school basketball team. H could move, move fast, and his ability to drop the balls into the net for three-point shots had helped the team be a strong contender for the State Championship which they had lost on the final elimination by only four points.

"Do you have your speech?" Samantha asked.

"Got it right here," Michael said as he patted the inside pocket of his suite.

"Do you think that everyone will be there?" he asked.

"Everyone was invited, but I don't know who will be able to come. Alex is still here in the city and said he would come, Henry was in London but was back for a few days, Aron's mom and dad could make it along with Erderlan and his wife. Larry Jr. and Beth will be coming up from Texas, but no one from Georgia could make it."

"It looks like we will need a half-dozen pups then?" Michael questioned.

"I've got the dog boxes ready," Aron said. "We'll load them up just before we leave."

"Last call," Samantha said. "Bathroom anybody?"

"I'm fine Mom. You know you don't really have to ask anymore," Michael responded.

"Old habit," Samantha replied. "In a way you will always be about six-years-old to me."

"Away then. Let's get gone," Aron suggested.

Thankfully, the traffic was light and the weather was clear with a nice cooling breeze coming off the lake. Parking was a bit of a challenge, but they managed to get into a parking deck next to the Husky's football field. Close rows of chairs had been arranged on the field to front a temporary stage which had been prefabricated, stored and re-erected for the occasion.

As Valedictorian of the Lakeside High class of 2040, Michael was one of the speakers sitting on the stage. After the introductory benediction and a statement by the Superintendent of Schools, the moment that Michael had been dreading had arrived. It was time for him to deliver his speech. He rose and walked to the podium. He gave an introductory Husky howl which was repeated by his classmates.

"At this time, in this place it is fitting to acknowledge the faculty, staff and my fellow students who have all contributed to this happy, apprehensive and regret-filled day when we are all together one last time before we began our life journeys.

"Critical events that shaped my life began a few days after my birth when my Jewish, Christian and Moslem family members forged lasting bonds when they were trapped by a relentless blizzard and shared an unexpected windfall discovered by a stray dog.

"Uncle Henry, a black man from the South Side became a noted Shakespearian actor, teacher and scholar. Uncle Erderlan, a Syrian refuge, became a doctor at University Hospital to fulfill a lifelong dream interrupted by war. His wife, Aunt Keje, hheads a refugee resettlement organization. Uncle Alex no works as an employee of a moving company, but owns a chain of movers throughout the country. My Goldfarb and Williams grandparents have been able to enjoy a richer retirement.

"All of this was the result of a stray dog who after giving birth and witnessing the departure of the people who had sheltered her unleashed a hoard of gold coins concealed in an antique bed. This bed had been made by by a member of the Goldfarb family in the Mid-1800s and traveled from France to England and ultimately to Detroit as a valued family touchstone.

"The subsequent kidnapping of my dad and his brother-in-law by Mexican smugglers and the repelling of a Mexican gang's attack on my grandparent's ranch in West Texas further cemented these relationships as Uncles Alex, Henry and Erderlan helped repel the attack and capture the gang. Jesus a former refuge from Guatemala became a sometimes family member and entrepreneur in Artificial Intelligence implementation.

"My life experiences illustrate how a family can be formed as a mutually supporting unit from those who are united in common purpose and circumstances even though they came from widely different backgrounds.

"Would those family members who are present please stand and face the audience?"

Each of the named family members stood and each grasped a struggling puppy under the front legs and presented them to the audience.

"Please give a Husky howl in honor of my family and those who have helped you on your life journeys."

With this statement a general howl erupted from the audience while Samantha and Aron looked at each other and smiled.

The End

Printed in the USA
CPSIA information can be obtained
at www.ICGtesting.com
LVHW091757200724
786053LV00037B/179